ITEM

Health and Safety
Executive

Controlling noise at work

The Control of Noise at Work
Regulations 2005

Guidance on Regulations

HSE Books

© *Crown copyright 2005*

First published 1998
Second edition 2005

ISBN 0 7176 6164 4

This guidance is issued by the Health and Safety Executive. Following the guidance is not compulsory and you are free to take other action. But if you do follow the guidance you will normally be doing enough to comply with the law. Health and safety inspectors seek to secure compliance with the law and may refer to this guidance as illustrating good practice.

Contents

Introduction

1   Hearing damage caused by exposure to noise at work is permanent and incurable. Research estimates that over 2 million people are exposed to noise levels at work that may be harmful. There are many new cases of people receiving compensation for hearing damage each year, through both civil claims and the Government disability benefit scheme, with considerable costs to industry, society and, most importantly, the people who suffer the disability.

2   Hearing loss is usually gradual due to prolonged exposure to noise. It may only be when damage caused by noise over the years combines with normal hearing loss due to ageing that people realise how deaf they have become. Hearing damage can also be caused immediately by sudden, extremely loud noises. Exposure to noise can also cause tinnitus, which is a sensation of noises in the ears such as ringing or buzzing. Tinnitus may occur in combination with hearing loss.

3   These conditions are entirely preventable if:

(a)   employers take action to reduce exposure to noise and provide personal hearing protection and health surveillance to employees;

(b)   manufacturers design tools and machinery to operate more quietly; and

(c)   employees make use of the personal hearing protection or other control measures supplied.

The Control of Noise at Work Regulations 2005

4   The Control of Noise at Work Regulations 2005 (the Noise Regulations) are based on a European Union Directive* requiring similar basic laws throughout the Union on protecting workers from the risks caused by noise. They do not apply to members of the public exposed to noise from their non-work activities, or making an informed choice to go to noisy places. They replace the Noise at Work Regulations 1989, which have been in force since 1990.

5   The duties in the Noise Regulations are in addition to the general duties set out in the Health and Safety at Work etc Act 1974 (the HSW Act). These general duties extend to the safeguarding of the health and safety of people who are not your employees, such as students, voluntary workers, visitors and members of the public. Employees also have duties under the HSW Act to take care of their own health and safety and that of others whom their work may affect; and to co-operate with employers so that they may comply with health and safety legislation.

6   The main differences from the 1989 Noise Regulations are:

(a)   the two action values for daily noise exposure have been reduced by 5 dB to 85 dB and 80 dB;

(b)   there are now two action values for peak noise at 135 dB and 137 dB;

(c)   there are new exposure limit values of 87 dB (daily exposure) and 140 dB (peak noise) which take into account the effect of wearing hearing protection and which must not be exceeded;

(d)   there is a specific requirement to provide health surveillance where there is a risk to health.

* Council and Parliament Directive 2003/10/EC of 6 February 2003 on the minimum health and safety requirements regarding the exposure of workers to the risks arising from physical agents (noise).

About this book

7   This book replaces the 1998 edition of L108 (ISBN 0 7176 1511 1). It is primarily aimed at employers. Part 1 includes the Control of Noise at Work Regulations 2005 together with guidance on what they mean. It sets out your legal obligations as an employer to control risks to workers' health and safety from noise.

8   Parts 2-6 include more detailed advice on how to assess risks, practical noise control, how to select and use hearing protection, what to consider when buying and hiring equipment, and how to develop health surveillance procedures.

9   Appendix 4 is for anyone involved in manufacturing or supplying machinery for use at work. It also refers to work equipment legislation as it applies to noise.

10   The other detailed Appendices are included for those who provide the employer with competent advice and services.

11   The Health and Safety Executive (HSE) has also published a free leaflet for employers INDG362(rev1)[1] which contains advice on how to comply with the Noise Regulations, as well as a pocket card for employees INDG363(rev1).[2] HSE also publishes guidance on noise control in specific industrial sectors and for particular types of machine. You can obtain information on these publications from HSE's Infoline (see 'Further information') or the HSE website (www.hse.gov.uk/noise).

12   HSE will periodically review this book. If you have any comments on it, please write to the Noise and Vibration Programme Unit, Health and Safety Executive, Rose Court, 2 Southwark Bridge, London SE1 9HS.

Managing noise risks

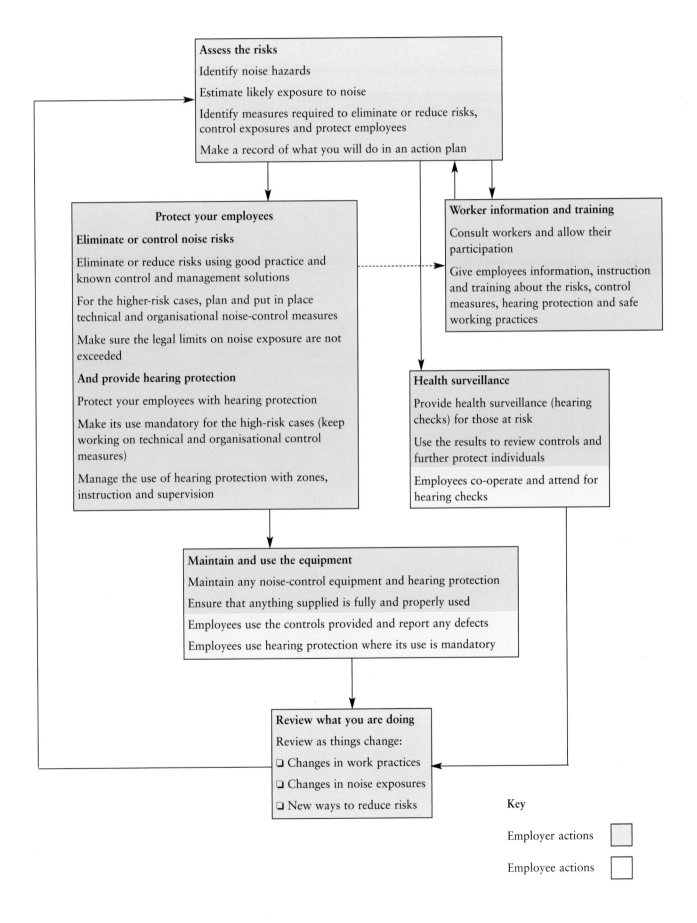

Assess the risks

Identify noise hazards

Estimate likely exposure to noise

Identify measures required to eliminate or reduce risks, control exposures and protect employees

Make a record of what you will do in an action plan

Protect your employees

Eliminate or control noise risks

Eliminate or reduce risks using good practice and known control and management solutions

For the higher-risk cases, plan and put in place technical and organisational noise-control measures

Make sure the legal limits on noise exposure are not exceeded

And provide hearing protection

Protect your employees with hearing protection

Make its use mandatory for the high-risk cases (keep working on technical and organisational control measures)

Manage the use of hearing protection with zones, instruction and supervision

Worker information and training

Consult workers and allow their participation

Give employees information, instruction and training about the risks, control measures, hearing protection and safe working practices

Health surveillance

Provide health surveillance (hearing checks) for those at risk

Use the results to review controls and further protect individuals

Employees co-operate and attend for hearing checks

Maintain and use the equipment

Maintain any noise-control equipment and hearing protection

Ensure that anything supplied is fully and properly used

Employees use the controls provided and report any defects

Employees use hearing protection where its use is mandatory

Review what you are doing

Review as things change:

❏ Changes in work practices

❏ Changes in noise exposures

❏ New ways to reduce risks

Key

Employer actions

Employee actions

PART 1: LEGAL DUTIES OF EMPLOYERS TO PREVENT DAMAGE TO HEARING

Regulation 1

Regulation

1

Citation and commencement

These Regulations may be cited as the Control of Noise at Work Regulations 2005 and shall come into force on 6th April 2006, except that –

(a) for the music and entertainment sectors only they shall not come into force until 6th April 2008; and

(b) subject to regulation 3(4), regulation 6(4) shall not come into force in relation to the master and crew of a seagoing ship until 6th April 2011.

Guidance

1

Transitional periods

13   Regulation 1(a) defers the application of the Noise Regulations in the 'music and entertainment' sectors until 6 April 2008. Until this date the Noise at Work Regulations 1989 will continue to apply (see regulation 15(3)). This two-year transitional period applies to all workplaces where live music is played or where recorded music is played in a restaurant, bar, public house, disco or nightclub, or alongside live music or a live dramatic or dance performance. Specific guidance on the practical measures that can be taken in these workplaces will be issued before the transitional period expires. If you are an employer in the music and entertainment sectors you should remember that under the 1989 Regulations there is a duty to reduce the risk of hearing damage to your employees to the lowest level reasonably practicable as well as other duties related to action levels – you still need to take action to protect workers in those sectors from risks from noise.

14   Regulation 1(b) defers until 6 April 2011 application of the exposure limit values only (regulation 6(4)) for the master and crew of a sea-going ship (also see paragraph 24).

Regulation 2

Regulation

2

Interpretation

(1)   In these Regulations –

"daily personal noise exposure" means the level of daily personal noise exposure of an employee as ascertained in accordance with Schedule 1 Part 1, taking account of the level of noise and the duration of exposure and covering all noise;

"emergency services" include –

(a)   police, fire, rescue and ambulance services;

(b)   Her Majesty's Coastguard;

"enforcing authority" means the Executive or local authority, determined in accordance with the provisions of the Health and Safety (Enforcing Authority) Regulations 1998;[a]

(a) SI 1998/494, as amended by SI 1999/3232, SI 1999/2024, SI 2002/2675 and SI 2004/3168.

Regulation

"the Executive" means the Health and Safety Executive;

"exposure limit value" means the level of daily or weekly personal noise exposure or of peak sound pressure set out in regulation 4 which must not be exceeded;

"health surveillance" means assessment of the state of health of an employee, as related to exposure to noise;

"lower exposure action value" means the lower of the two levels of daily or weekly personal noise exposure or of peak sound pressure set out in regulation 4 which, if reached or exceeded, require specified action to be taken to reduce risk;

"the music and entertainment sectors" mean all workplaces where –

(a) live music is played; or

(b) recorded music is played in a restaurant, bar, public house, discotheque or nightclub, or alongside live music or a live dramatic or dance performance;

"noise" means any audible sound;

"peak sound pressure" means the maximum sound pressure to which an employee is exposed, ascertained in accordance with Schedule 2;

"risk assessment" means the assessment of risk required by regulation 5;

"upper exposure action value" means the higher of the two levels of daily or weekly personal noise exposure or of peak sound pressure set out in regulation 4 which, if reached or exceeded, require specified action to be taken to reduce risk;

"weekly personal noise exposure" means the level of weekly personal noise exposure as ascertained in accordance with Schedule 1 Part 2, taking account of the level of noise and the duration of exposure and covering all noise; and

"working day" means a daily working period, irrespective of the time of day when it begins or ends, and of whether it begins or ends on the same calendar day.

(2) In these Regulations, a reference to an employee being exposed to noise is a reference to the exposure of that employee to noise which arises while he is at work, or arises out of or in connection with his work.

2

Regulation 3

Regulation

Application

(1) These Regulations shall have effect with a view to protecting persons against risk to their health and safety arising from exposure to noise at work.

(2) Where a duty is placed by these Regulations on an employer in respect of his employees, the employer shall, so far as is reasonably practicable, be under a like duty in respect of any other person at work who may be affected by the work carried out by the employer except that the duties of the employer –

(a) under regulation 9 (health surveillance) shall not extend to persons who are not his employees; and

3

Regulation

3

(b)   under regulation 10 (information, instruction and training) shall not extend to persons who are not his employees, unless those persons are present at the workplace where the work is being carried out.

(3)   These Regulations shall apply to a self-employed person as they apply to an employer and an employee and as if that self-employed person were both an employer and an employee, except that regulation 9 shall not apply to a self-employed person.

(4)   These Regulations shall not apply to the master or crew of a ship or to the employer of such persons in respect of the normal shipboard activities of a ship's crew which are carried out solely by the crew under the direction of the master, and for the purposes of this paragraph "ship" includes every description of vessel used in navigation, other than a ship forming part of Her Majesty's Navy.

Guidance

Purpose

15   The Noise Regulations are designed to protect against risks to both health and safety from exposure to noise – the health risk of hearing damage in those exposed, and safety risks such as the noise affecting the ability to hear instructions or warnings.

People who are not your employees

16   Sometimes your activities may cause employees of other employers to be exposed to noise, eg where contractors take noisy tools into quiet premises to do their job, or they go to do a quiet job in premises that are already noisy. Regulation 3(2) places duties on all the employers involved and each will have a responsibility:

(a)   to their own employees; and

(b)   so far as is reasonably practicable, to any other person at work who is affected by the work they do.

17   This responsibility applies to all the duties under the Noise Regulations except health surveillance (regulation 9), which you do not have to provide for anyone other than your own employees, while you only need to provide information, instruction and training (regulation 10) to the employees of others in relation to the specific job they are doing for you.

18   In most cases employers will need to exchange information and collaborate to ensure they fulfil their duties without confusion or unnecessary duplication. On multi-contractor sites they will usually need to agree on who is to co-ordinate action to comply with health and safety requirements; this will normally be the person in overall control of the work. This person should make sure that responsibilities for controlling risks are clearly defined. For example it will often be appropriate for the employer in overall control to make sure that risks are assessed and that the information on noise is made available to all affected employers, while the actual employer of each worker provides any training needed.

19   Where contractors and sub-contractors are involved it is usually best for responsibilities to be set out in the contractual arrangements. For construction projects, the principal contractor under the Construction (Design and Management) Regulations 1994 (as amended)[3] should ensure co-operation between all contractors.

20   If you are in charge of premises you should make sure that visiting workers, including contractors, know in which areas they should use hearing protection and know how to obtain it. You may wish to include this information in induction information for new staff and/or in general training.

21   If your employees need to visit premises controlled by someone else (eg for maintenance or survey work) you will need to consider whether exposure over the exposure action values is likely, and what can reasonably be done to control it (eg by providing hearing protection adequate for the worst likely exposure). Employees should co-operate with their employers so far as this is necessary so that employers can meet their obligations.

The self-employed

22   Regulation 3(3) defines both employer and employee to include self-employed people. So if you are self-employed you will need to take action as set out in the Noise Regulations to protect yourself from noise risks. Although self-employed people are not required to provide themselves with health surveillance in accordance with regulation 9, it is recommended that they follow the guidance in Part 6 and, where appropriate, consult an occupational health service provider. This will ensure that early signs of hearing loss are identified and will allow risks to be reviewed and revised as necessary.

Trainees

23   The Health and Safety (Training for Employment) Regulations 1990[4] require trainees on relevant work training schemes in the workplace (but not those on courses at educational establishments such as universities or schools) to be treated as the employee of the person whose undertaking is providing the training. Your duties towards trainees will include all the requirements of the Noise Regulations including assessment and control of risks, provision of health surveillance, provision of information, instruction and training and consideration of whether any trainees might be at particular risk.

Application to ships, other vessels and aircraft

24   Regulation 3(4) states that the Noise Regulations do not apply to the master and crew of a ship. This refers to work done by the crew under the control of the ship's master when the ship is under way or work done by them in harbour when no shore-based workers are involved. However, in the future, similar regulations administered by the Maritime and Coastguard Agency will apply to all vessels in UK waters and to UK-registered vessels in international waters. The Noise Regulations do apply to work taking place in ships, boats and other vessels operated by Her Majesty's Navy and to work on any vessel carried out alongside shore workers when it is moored or in dock.

25   The Noise Regulations apply to aircraft in flight over British soil.

Regulation 4

Exposure limit values and action values

(1)   The lower exposure action values are –

(a)   a daily or weekly personal noise exposure of 80 dB (A-weighted); and

(b)   a peak sound pressure of 135 dB (C-weighted).

(2)   The upper exposure action values are –

(a)   a daily or weekly personal noise exposure of 85 dB (A-weighted); and

(b)   a peak sound pressure of 137 dB (C-weighted).

(3)   The exposure limit values are –

(a)   a daily or weekly personal noise exposure of 87 dB (A-weighted); and

(b)   a peak sound pressure of 140 dB (C-weighted).

(4)   Where the exposure of an employee to noise varies markedly from day to day, an employer may use weekly personal noise exposure in place of daily personal noise exposure for the purpose of compliance with these Regulations.

(5)   In applying the exposure limit values in paragraph (3), but not in applying the lower and upper exposure action values in paragraphs (1) and (2), account shall be taken of the protection given to the employee by any personal hearing protectors provided by the employer in accordance with regulation 7(2).

26   The exposure action values are the levels of exposure to noise at which you are required to take certain actions (see regulations 5, 6, 7, 9 and 10). The exposure limit values are the levels of noise above which an employee may not be exposed (see regulation 6(4)). Your risk assessment (see regulation 5) should include an assessment of the likely noise exposure of your employees for comparison with these exposure action and exposure limit values. A formula for calculating daily exposure is shown in Schedule 1 Part 1 and for peak sound pressure in Schedule 2 to the Noise Regulations.

Weekly exposure

27   Regulation 4(4) allows you to calculate exposures over a week rather than over a day in circumstances where noise exposure varies markedly from day to day. The formula for calculating weekly exposure is shown in Schedule 1 Part 2 to the Noise Regulations.

28   Use of weekly exposure might be appropriate in situations where noise exposure varies markedly from day to day, eg where people use noisy power tools on one day in the week but not on others. It is only likely to be appropriate where daily noise exposure on one or two working days in a week is at least 5 dB higher than the other days, or the working week comprises three or fewer days of exposure.

29   When considering whether to use weekly averaging it is important to:

(a)   ensure there is no increase in risk to health. It would not, for example, be acceptable to expose workers to very high noise levels on a single day without providing them with hearing protection. There is an overriding

requirement to reduce risk to as low a level as is reasonably practicable (regulation 6(1));

(b) consult the workers concerned and their safety or employee representatives on whether weekly averaging is appropriate;

(c) explain to workers the purpose and possible effects of weekly averaging.

Taking account of hearing protection

30 Note that the requirements in the Noise Regulations relating to assessment of risk and exposure, actions to take to reduce risk and exposure, and the levels at which they are required to be taken, **do not** allow you to take account of the reduction of noise provided by wearing hearing protection.

31 Regulation 4(5) explains that the only exception when this reduction can be taken into account is in relation to the exposure limit values (regulation 6(4)). The reduction is likely to be an estimate based on the information provided by the manufacturer of the particular hearing protection device used (see Appendix 3 for further information). Whether the exposure limit values are complied with will depend, not only on the reduction provided by the hearing protectors, but also on whether the hearing protection is in good working order, is appropriate for the type of noise and is properly worn.

4

Regulation 5

Assessment of the risk to health and safety created by exposure to noise at the workplace

Regulation

(1) *An employer who carries out work which is liable to expose any employees to noise at or above a lower exposure action value shall make a suitable and sufficient assessment of the risk from that noise to the health and safety of those employees, and the risk assessment shall identify the measures which need to be taken to meet the requirements of these Regulations.*

(2) *In conducting the risk assessment, the employer shall assess the levels of noise to which workers are exposed by means of –*

(a) *observation of specific working practices;*

(b) *reference to relevant information on the probable levels of noise corresponding to any equipment used in the particular working conditions; and*

(c) *if necessary, measurement of the level of noise to which his employees are likely to be exposed,*

and the employer shall assess whether any employees are likely to be exposed to noise at or above a lower exposure action value, an upper exposure action value, or an exposure limit value.

(3) *The risk assessment shall include consideration of –*

(a) *the level, type and duration of exposure, including any exposure to peak sound pressure;*

(b) *the effects of exposure to noise on employees or groups of employees whose health is at particular risk from such exposure;*

5

(c) as far as is practicable, any effects on the health and safety of employees resulting from the interaction between noise and the use of ototoxic substances at work, or between noise and vibration;

(d) any indirect effects on the health and safety of employees resulting from the interaction between noise and audible warning signals or other sounds that need to be audible in order to reduce risk at work;

(e) any information provided by the manufacturers of work equipment;

(f) the availability of alternative equipment designed to reduce the emission of noise;

(g) any extension of exposure to noise at the workplace beyond normal working hours, including exposure in rest facilities supervised by the employer;

(h) appropriate information obtained following health surveillance, including, where possible, published information; and

(i) the availability of personal hearing protectors with adequate attenuation characteristics.

(4) The risk assessment shall be reviewed regularly, and forthwith if –

(a) there is reason to suspect that the risk assessment is no longer valid; or

(b) there has been a significant change in the work to which the assessment relates,

and where, as a result of the review, changes to the risk assessment are required, those changes shall be made.

(5) The employees concerned or their representatives shall be consulted on the assessment of risk under the provisions of this regulation.

(6) The employer shall record –

(a) the significant findings of the risk assessment as soon as is practicable after the risk assessment is made or changed; and

(b) the measures which he has taken and which he intends to take to meet the requirements of regulations 6, 7 and 10.

5

Risk assessment

32   The purpose of the risk assessment is to enable you as the employer to make a valid decision about whether your employees are at risk from exposure to noise and what action may be necessary to prevent or adequately control that exposure. It enables you to demonstrate readily to others who have an interest, eg safety representatives and enforcement authorities, that you have, from the earliest opportunity, considered:

(a) all the factors related to the risks from noise exposure;

(b) the steps which need to be taken to achieve and maintain adequate control of the risks;

5

(c)   the need for health surveillance;

(d)   how to put the steps you have decided on into action.

33   The risk assessment must take into account all noise exposure at work, including, for example, piped music and personal stereos.

When is a noise risk assessment needed?

34   You must do a risk assessment if any employee is likely to be exposed to noise at or above the lower exposure action values. A person's daily noise exposure depends on both noise level and length of exposure.

35   If your workplace is intrinsically noisy, ie it is significantly noisier than you would expect from the sounds of everyday life, it is possible that the noise levels will exceed 80 dB. This is comparable to the noise level of a busy street, a typical vacuum cleaner or a crowded restaurant – you will be able to hold a conversation, but the noise will be intrusive. Working in an environment of 80 dB for eight hours will result in exposure at the lower exposure action value.

36   To get a rough estimate of whether a risk assessment is required you could use the simple tests in Table 1.

Test	Probable noise level	A risk assessment will be needed if the noise is like this for more than:
The noise is intrusive but normal conversation is possible (see paragraph 35)	80 dB	6 hours
You have to shout to talk to someone 2 m away	85 dB	2 hours
You have to shout to talk to someone 1 m away	90 dB	45 minutes

Table 1 Simple tests to see if a noise risk assessment is needed

37   For peak noise, some sources which may typically lead to exposure above the lower exposure action value are explosive sources, impactive tools, drop forges, punch presses and firearms. More advice on peak noise exposure is given in Appendix 2.

38   Deciding whether you need to do a noise risk assessment should not be time-consuming. If you are in any doubt, it would be best to assume that the lower exposure action values have been exceeded. If you are satisfied that your employees are not exposed at or above the lower exposure action values it is sufficient to record that fact. No further action will be necessary except to ensure that noise exposures are not increased and to take action if they are.

39   See Part 2 for more detailed guidance on risk assessment. Paragraphs 40-63 explain what various terms in regulation 5 mean and their effects on the risk assessment.

Guidance

5(1)

"A suitable and sufficient assessment"

40   An assessment will be suitable and sufficient if it:

(a)   has been drawn up by someone who is competent to carry out the task;

(b)   is based on advice and information from competent sources;

(c)   identifies where there may be a risk from noise and who is likely to be affected;

(d)   contains a reliable estimate of your employees' noise exposures and a comparison of exposure with the exposure action values and limit values;

(e)   identifies the measures necessary to eliminate risks and exposures or reduce them to as low a level as is reasonably practicable;

(f)   identifies those employees who need to be provided with health surveillance and whether any employees are at particular risk.

"The measures which need to be taken"

41   When assessing the work processes which expose your employees to noise you should think about what needs to be done to eliminate or at least reduce the risks, and draft a plan of action. If exposure is likely to be at or above the upper exposure action values, you must establish a formal programme of control measures (see regulation 6(2)). Further guidance on what to do is in Parts 2 and 3.

Guidance

5(2)

"Shall assess the levels of noise"

42   Your risk assessment must contain an assessment of the noise levels to which your employees are exposed, for comparison with the exposure action values. Where exposure varies from day to day you will need to assess the various daily exposures, taking into account both a typical day and a worst-case day. More detailed advice on assessing noise exposure is given in Part 2.

43   You are not required to make a highly precise or definitive assessment of individual employees' noise exposure, such as would be obtained by making detailed measurements. Your assessment of exposure must be a reliable estimate with sufficient precision for you to be able to show whether exposure action values are likely to be exceeded. Your assessment of exposure will only be reliable if it uses data which is reasonably representative of individuals' exposure. You would be expected to use data from measurements of noise where other sources cannot give you reliable and representative data.

44   Uncertainties in an assessment of exposure to noise can arise from variability in the level of noise and in the duration of exposure. If you assess exposure as being close to an exposure action value then you should proceed as if the exposure action value has been exceeded, or ensure that your assessment is sufficiently precise to demonstrate that exposure is below the exposure action value.

Guidance

5(2)(a)

"Observation of specific working practices"

45   To assess noise exposure you need to understand the work your employees do and how they do it. Workers may not be exposed to the same noise levels throughout the day, and they may only spend part of their time in noisy areas. Your employees may not do their work in the way you assume or expect. There may be local practices which differ from what is usual in your industry or sector.

So work patterns, work tasks and work practices all need careful consideration. To take these factors into account you will need to observe the employees' work patterns and practices during representative periods.

"Relevant information on the probable levels of noise"

46   Any information you use to estimate noise levels needs to match as closely as possible the conditions and practices in your workplace. This is particularly important where you are using data which has not resulted from measurements in your workplace. So if you consult published information on typical noise levels in certain industries, or use noise data from machinery manufacturers, you need to be sure the data can be taken to be representative of your work. More advice on the use of manufacturers' data is in Part 4.

"Measurement of the level of noise"

47   The Noise Regulations require you to make measurements of noise 'if necessary'. Measurements will be necessary if you cannot make a reliable estimate of your employees' exposure in other ways. You may also wish to use measurements to demonstrate that the noise exposure is below a particular value so that you can assure yourself and others that you are complying with the Noise Regulations, and if you require confirmation that your control actions have reduced exposure.

48   You should ensure that any measurements are carried out by someone who is competent, ie someone who has the relevant skills, knowledge and experience to undertake measurements in your particular working environment. More detailed advice on measuring noise in the workplace is in Appendix 1.

Factors to consider in assessing risks

"Level, type and duration of exposure"

49   The factors which govern a person's daily noise exposure are the level of noise and the length of time they are exposed to it. The greater the noise level or the longer the duration of exposure, the greater the person's noise exposure will be. Some methods to help you estimate likely noise exposure by combining noise levels and time of exposure are in Part 2. Other characteristics of the noise, such as the frequency and whether the noise is continuous or characterised by high levels of short duration, may also affect the risk.

"Employees or groups of employees whose health is at particular risk"

50   Some workers should be given particular consideration within your risk assessment, eg people with a pre-existing hearing condition, those with a family history of deafness (if known), pregnant women and young people. The Management of Health and Safety at Work Regulations 1999[5] prohibit you from employing anyone under 18 where there is a risk to health from noise.

"Any effects . . . from the interaction between noise and the use of ototoxic substances at work, or between noise and vibration"

51   Some studies have suggested that there is a link between exposure to hand-transmitted vibration and hearing loss, meaning that workers may be more vulnerable to noise-induced hearing loss if they are exposed to hand-transmitted vibration. Other studies have suggested that some chemicals, particularly solvents, can act in combination with noise to cause further damage to hearing than the

Guidance

5(3)(c)

noise or chemical exposures alone. Where there are likely to be such mixed exposures in your workplace you should note this within your risk assessment and monitor developments on these issues. If you suspect the use of chemicals or vibrating equipment might increase the risk of hearing damage to any of your employees, you could:

(a) consider whether you can limit their exposure by reducing the time spent on particular tasks;

(b) monitor the health surveillance results of those workers;

(c) increase the frequency of health surveillance for those workers.

Guidance

5(3)(d)

"Indirect effects . . . resulting from the interaction between noise and audible warning signals"

52   Noise can mask important warning signals and messages, leading to potential safety issues. You will need to consider the characteristics of any audible warning and information signals in your workplace to take account of the possible masking effects of the general noise environment and of any hearing protection worn. You may want to consider visual warnings, or the use of hearing protection with communication facilities (see Part 5).

Guidance

5(3)(e)

"Information provided by the manufacturers of work equipment"

53   Suppliers of machinery are legally required to provide information on the noise emission from their machinery if it exceeds a certain level (see Appendix 4). They must tell you what the noise level is likely to be at the operator position if it exceeds an A-weighted sound pressure of 70 dB (or a C-weighted peak sound pressure of 130 dB). They must also tell you what the total noise emitted by the machine is (the 'sound power level') if the operator position noise level exceeds 85 dB. It may be possible to use this information in your risk assessment and to assess noise exposure if the information is relevant to your work.

Guidance

5(3)(f)

"The availability of alternative equipment designed to reduce the emission of noise"

54   Manufacturers are legally required to ensure that machinery is designed and constructed to reduce risks from noise to the lowest level taking account of technical progress. You should expect to be able to find equipment on the market which reflects technical progress in low noise emissions, and for progress to continue over time. In deciding whether you have done enough to reduce risks from noise you will need to consider whether there are lower-noise alternatives to the tools and machinery you are using.

55   The implementation of a positive purchasing policy in relation to noise is covered in paragraphs 72-74. Part 4 gives more detailed information on how to select quieter tools and machines, and Appendix 4 on the legal duties of manufacturers and suppliers.

Guidance

5(3)(g)

"Extension of exposure to noise at the workplace beyond normal working hours"

56   Employees can also be exposed to noise at the workplace over and above their normal working hours, eg during overtime, extended shifts, lunch breaks or rest times. You need to take this exposure into account in the risk assessment. If you provide rest facilities on site, you must ensure it is quiet enough inside them so that people can rest. For example, in sleeping quarters, noise levels over 45 dB are known to cause annoyance and/or sleep disturbance. Minimising exposure

during rest periods and breaks is particularly important to allow employees some respite from the noise.

"Appropriate information obtained following health surveillance"

57   You should arrange to receive health surveillance data (anonymised and grouped to protect medical-in-confidence information about individual workers) relating to your own business. This will indicate whether new cases of noise-induced hearing loss are developing or whether existing cases have worsened. This will help you decide whether the risk is being controlled effectively and whether you need to do more to control it.

58   General data relating to the results of health surveillance in your particular sector or industry may also provide useful information. This information may be published or made available by HSE, trade associations, industry-specific journals or other publications.

"The availability of personal hearing protectors with adequate attenuation characteristics"

59   You will need to consider whether suitable hearing protection is available for reducing noise exposure in the particular circumstances of your work. The hearing protection has to be appropriate to the level and character of the noise and to reduce the noise adequately. You need to take account of this when planning measures to control the risk and you need to consider developments in hearing protection technology and design when reviewing those measures. Hearing protection should not be used as an alternative to controlling noise by technical and organisational means, but it can be used as an interim measure while these other controls are being developed. Further guidance on the use of hearing protection is given in paragraphs 82-87 and in Part 5.

"Reviewed regularly"

60   The risk assessment should be reviewed and updated when circumstances change in your workplace which might alter the level of exposure or where there are technological changes or changes to the availability, applicability or cost of noise-control measures. The review of the assessment should be part of an ongoing noise-risk management and control programme which can pick up changes as they occur. Even if you consider there have been no changes, you should review your assessment at least every two years. See Part 2 paragraphs 181-183 for more information.

"The employees concerned or their representatives shall be consulted"

61   It is important to talk to the workers concerned and their employee or safety representatives, not only to tell them what you are doing, but also to seek their advice, help and co-operation on what is achievable and practical. They can provide valuable advice on how measures you propose to take will affect their work and may suggest action you can take which you had not considered.

"Record the significant findings of the risk assessment . . . and the measures which he has taken"

62   You must make a record of your noise risk assessment covering:

(a)   the major findings, including which of your employees are at risk, the level of risk and exposure, and under what circumstances the risks occur;

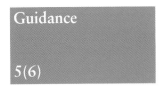

(b)   the action you have taken or intend to take, with timescales and allocation of responsibility.

63   Further information on what your record should cover is in Part 2 paragraph 180.

Guidance

5

Competence

64   To carry out the tasks which may be involved in noise risk assessment properly requires competence in particular areas, eg drawing up the risk assessment (paragraph 40), measuring noise exposure (paragraphs 47-48) and assessing the likely effectiveness of control measures. Regulation 7 of the Management of Health and Safety at Work Regulations 1999 requires the employer to have access to competent help in applying health and safety law. You may have some of the necessary competencies yourself. You may have people within your workforce who are competent to carry out some of the tasks or provide you with the necessary information, perhaps with some training required. Alternatively, you may find that you need to call in competent people from outside your company, such as consultants, to carry out some or all of the work. More details are given in Part 2 paragraphs 184-187.

Regulation 6

Elimination or control of exposure to noise at the workplace

Regulation

6

(1)   The employer shall ensure that risk from the exposure of his employees to noise is either eliminated at source or, where this is not reasonably practicable, reduced to as low a level as is reasonably practicable.

(2)   If any employee is likely to be exposed to noise at or above an upper exposure action value, the employer shall reduce exposure to as low a level as is reasonably practicable by establishing and implementing a programme of organisational and technical measures, excluding the provision of personal hearing protectors, which is appropriate to the activity.

(3)   The actions taken by the employer in compliance with paragraphs (1) and (2) shall be based on the general principles of prevention set out in Schedule 1 to the Management of Health and Safety Regulations 1999[(a)] and shall include consideration of –

(a)   other working methods which reduce exposure to noise;

(b)   choice of appropriate work equipment emitting the least possible noise, taking account of the work to be done;

(c)   the design and layout of workplaces, work stations and rest facilities;

(d)   suitable and sufficient information and training for employees, such that work equipment may be used correctly, in order to minimise their exposure to noise;

(e)   reduction of noise by technical means;

(f)   appropriate maintenance programmes for work equipment, the workplace and workplace systems;

(a) SI 1999/3242, as amended by SI 2003/2457.

(g)   *limitation of the duration and intensity of exposure to noise; and*

(h)   *appropriate work schedules with adequate rest periods.*

(4)   *The employer shall –*

(a)   *ensure that his employees are not exposed to noise above an exposure limit value; or*

(b)   *if an exposure limit value is exceeded forthwith –*

    (i)   *reduce exposure to noise to below the exposure limit value;*

    (ii) *identify the reason for that exposure limit value being exceeded; and*

    (iii) *modify the organisational and technical measures taken in accordance with paragraphs (1) and (2) and regulations 7 and 8(1) to prevent it being exceeded again.*

(5)   *Where rest facilities are made available to employees, the employer shall ensure that exposure to noise in these facilities is reduced to a level suitable for their purpose and conditions of use.*

(6)   *The employer shall adapt any measure taken in compliance with the requirements of this regulation to take account of any employee or group of employees whose health is likely to be particularly at risk from exposure to noise.*

(7)   *The employees concerned or their representatives shall be consulted on the measures to be taken to meet the requirements of this regulation.*

6

Controlling noise

65   This regulation places duties on you:

(a)   to take action to eliminate risks from noise exposure completely wherever it is reasonably practicable to do so (regulation 6(1));

(b)   if it is not reasonably practicable to eliminate the risks completely, to reduce them to as low a level as is reasonably practicable (regulation 6(1));

(c)   to introduce a formal programme of measures to reduce noise exposure whenever an employee's exposure to noise is likely to exceed the upper exposure action values (these measures **cannot** include hearing protection, which is addressed separately) (regulation 6(2));

(d)   not to expose anyone above the exposure limit values (regulation 6(4)) (see paragraphs 78-80).

6

Eliminate or reduce risk

66   This general duty applies whenever there is a risk from noise and irrespective of whether any exposure action values are exceeded. It is likely, however, that only inexpensive and simple measures will be reasonably practicable if the lower exposure action values are not exceeded. To comply with this duty you need to:

(a)   consider whether there are alternative processes, equipment and/or working methods which would eliminate risks from noise exposure;

6(1)

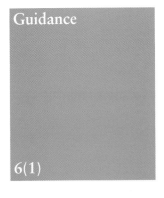

(b)   follow good practice and industry standard control measures (see Part 2 paragraphs 169-171);

(c)   take noise into account when selecting tools and machinery (see paragraphs 72-74);

(d)   maintain machinery in accordance with manufacturers' recommendations (see paragraph 75);

(e)   explore any opportunity to provide your employees with periods of relief from noise exposure (see paragraph 77).

"Establishing and implementing a programme of organisational and technical measures"

67   The action plan produced during your noise risk assessment should describe a programme of control measures and your plans to put it into action with realistic timescales. The programme of control measures should be devised to reduce noise exposures so far as is reasonably practicable.

68   The actions you take will depend on the particular work activities and processes and the possibilities for control, but in general you should:

(a)   identify what is possible to control noise exposures, how much reduction could be achieved and so what is reasonably practicable;

(b)   establish priorities for action and a timetable;

(c)   assign responsibilities to individuals to deliver the various parts of the programme;

(d)   ensure that the work involved in implementing the noise-control measures is carried out;

(e)   check that what you have done has been effective in reducing noise exposures.

69   Some controls may take time to put in place, particularly where equipment must be replaced or new industrial processes developed. Other controls may be considered to be not reasonably practicable but may become so over time as circumstances change. You will need regularly to review the feasibility of further noise reductions.

"The general principles of prevention"

70   In identifying and putting in place appropriate noise-control and risk-reduction measures you should follow the general principles of prevention set out in Schedule 1 to the Management of Health and Safety at Work Regulations 1999:

(a)   avoiding risks;

(b)   evaluating the risks which cannot be avoided;

(c)   combating the risks at source;

(d)   adapting the work to the individual, especially as regards the design of workplaces, the choice of work equipment and the choice of working and production methods, with a view, in particular, to alleviating monotonous work and work at a predetermined work-rate and to reducing their effect on health;

(e)   adapting to technical progress;

(f)   replacing the dangerous by the non-dangerous or the less dangerous;

(g)   developing a coherent overall prevention policy which covers technology, organisation of work, working conditions, social relationships and the influence of factors relating to the working environment;

(h)   giving collective protective measures priority over individual protective measures;

(i)   giving appropriate instructions to employees.

Guidance

6(3)

71   Regulation 6(3) lists several possible noise-control and risk-reduction methods, following the general principles of prevention. There are other ways of reducing noise and no single technique will be appropriate for every situation. A programme of noise control should adopt a systematic approach to identifying what can be done, and should not be restricted to considering what is listed in regulation 6(3). Part 3 gives practical advice on noise control, while paragraphs 72-77 include some basic guidance on some of the measures in regulation 6(3).

Guidance

6(3)(b)

"Choice of appropriate work equipment emitting the least possible noise"

72   For many types of equipment there will be models designed to be less noisy. Noise-reduction programmes are only likely to be effective if they include a positive purchasing policy which makes sure you take noise into account when selecting machinery. When buying, hiring or replacing equipment you should ask potential suppliers for information on the noise emission of machines under the conditions you intend to use it, and use that information to compare machines.

73   Where you find it is necessary to purchase machinery which causes workers to be exposed over the action levels, you will find that keeping a record of the reasons for the decision will help guide future action, eg by providing those responsible for future machine specifications with information on improvements that are needed.

74   Part 3 paragraphs 201-202 has more information about a positive purchasing policy and Part 4 has more information on the selection of quieter tools and machinery, including the use and limitations of manufacturers' noise data.

Guidance

6(3)(f)

"Appropriate maintenance programmes for work equipment"

75   Maintenance of machinery, carried out in accordance with the manufacturer's recommendations, can prevent noise emissions increasing over time. You should ensure that appropriate maintenance is performed on equipment so that its performance does not deteriorate to the extent that it puts employees at risk due to the noise emitted. Operators should be instructed to report any unusually high noise levels and check that machines are operating properly.

Guidance

6(3)(g)

"Limitation of the duration and intensity of exposure"

76   When all reasonably practicable steps have been taken to reduce noise levels the next step to reduce exposure is to limit its duration. The exposure points system described in Part 2 paragraphs 154-158 can be a useful management tool for this purpose.

Guidance

6(3)(h)

"Appropriate work schedules with adequate rest periods"

77 Workers exposed to loud noise should have the opportunity to spend time away from the noisy environment and, wherever possible, breaks should be taken in quiet zones. Even if this does not significantly reduce daily exposure it will help by allowing recuperation and, in some circumstances, preventing the need to wear hearing protection continuously.

Guidance

6(4)

Reduction of exposure below the exposure limit values

78 You must not permit an employee to be exposed above the exposure limit values. You will need to check whether your programme of control measures, including, in this case, the provision of hearing protection, is enough to prevent this level of exposure.

79 If you discover that an exposure limit value is exceeded, you must immediately take action to reduce exposure. Address the reasons for the overexposure by reviewing your programme of control measures. You should consider the technical and organisational controls, the adequacy of any hearing protection supplied and the systems you have in place to ensure that noise-control measures and hearing protection are fully and properly used and maintained.

80 You should not consider the exposure limit values to be a target for your noise control programme – remember that regulations 6(1) and 6(2) require you to reduce risks and exposures to as low a level as is reasonably practicable.

Guidance

6(6)

"Any employee or group of employees whose health is likely to be at particular risk from exposure to noise"

81 Paragraph 50 describes employees in this category. As well as special efforts to restrict exposure for such individuals an increased level of health surveillance may also be appropriate.

Regulation 7

Hearing protection

Regulation

7

(1) Without prejudice to the provisions of regulation 6, an employer who carries out work which is likely to expose any employees to noise at or above a lower exposure action value shall make personal hearing protectors available upon request to any employee who is so exposed.

(2) Without prejudice to the provisions of regulation 6, if an employer is unable by other means to reduce the levels of noise to which an employee is likely to be exposed to below an upper exposure action value, he shall provide personal hearing protectors to any employee who is so exposed.

(3) If in any area of the workplace under the control of the employer an employee is likely to be exposed to noise at or above an upper exposure action value for any reason the employer shall ensure that –

(a) the area is designated a Hearing Protection Zone;

(b) the area is demarcated and identified by means of the sign specified for the purpose of indicating that ear protection must be worn in paragraph 3.3 of Part II of Schedule 1 to the Health and Safety (Safety Signs and Signals) Regulations 1996;[a] and –

(a) SI 1996/341.

(c)   access to the area is restricted where this is practicable and the risk from exposure justifies it,

and shall ensure so far as is reasonably practicable that no employee enters that area unless that employee is wearing personal hearing protectors.

(4)   Any personal hearing protectors made available or provided under paragraphs (1) or (2) of this regulation shall be selected by the employer –

(a)   so as to eliminate the risk to hearing or to reduce the risk to as low a level as is reasonably practicable; and

(b)   after consultation with the employees concerned or their representatives.

The need for hearing protectors

82   Personal hearing protection should only be used:

(a)   where there is a need to provide additional protection beyond what has been achieved through noise-control measures under regulation 6;

(b)   as an interim measure while you are developing those control measures.

83   It should not be used as an alternative to controlling noise by technical and organisational means.

84   The duty to provide hearing protectors depends on the exposure levels:

(a)   Where employees are exposed between the lower and upper exposure action values you have to provide protectors to employees who ask for them but the Noise Regulations do not make their use compulsory.

(b)   Where employees are likely to be exposed at or above the upper exposure action values, you have to provide hearing protectors. Regulation 8 requires you to ensure the hearing protectors are used and requires your employees to use them. Under regulation 10 you will need to provide information to your employees about the protectors and how to obtain and use them.

85   Making the use of hearing protection compulsory for workers exposed below the upper exposure action values should be avoided, except within hearing protection zones.

86   Where workers are exposed above the upper exposure action values and are therefore required to wear hearing protection, you should not necessarily make it compulsory at all times throughout the working day, eg in areas or at times when noise levels are low. Hearing protection use should be targeted at particular noisy jobs and activities and be selected to reduce exposure at least to below the upper exposure action values.

87   Advice on the selection of suitable hearing protectors and their use, care and maintenance is in Part 5.

Hearing protection zones

88   Hearing protection zones provide a way for you to manage the use of hearing protection. They give a reminder to those employees for whom hearing protection is compulsory during particular jobs or activities. They also provide a

way of ensuring that employees or other people affected by the noise from those jobs or activities are protected.

89   You should designate as hearing protection zones any areas of your workplace where work is going on during which particular employees must be provided with, and use, hearing protection (regulations 7(2) and 8(1)(a)). You should also designate as hearing protection zones any areas of your workplace where the upper exposure action values would be likely to be exceeded if personnel spent a significant portion of the working day within them, even if access is generally infrequent, eg plant rooms or compressor houses.

90   Hearing protection zones can be fixed locations or be mobile, and can be permanent or temporary, depending on the nature of the activities and the source of noise.

91   You will need to ensure that no person enters a hearing protection zone unless it is necessary to carry out their work. Before entering a hearing protection zone people must put on suitable hearing protection and must wear it all the time they are within the zone. You should instruct employees and other people of these requirements, and put a system of supervision in place to ensure these instructions are followed.

92   You will need to mark hearing protection zones with signs showing that they are areas where hearing protection is needed. You should locate these signs at all entrances to the zones and at appropriate places within the zones as necessary. The sign need not include any words, but where wording is included it should convey the same meaning as the sign. Signs introduced under the Noise at Work Regulations 1989 referring to 'ear protection zones' are acceptable and need not be changed.

Figure 1 Hearing protection zone sign

93   The boundaries of hearing protection zones should be considered carefully. You should avoid hearing protection zones overlapping with designated or commonly used walkways. Zones should not extend any further than is necessary to protect people carrying out their normal work or any foreseeable non-typical tasks.

94   In situations where the boundaries of the zone cannot be marked, eg where the work requires people to move the noise sources about a great deal, you should make adequate alternative arrangements to help make sure that people know where or when protectors should be worn. These could include:

(a)   attaching signs to tools warning that people who are using them must wear hearing protectors;

(b)   written and verbal instructions on how to recognise where and when protectors should be worn, eg by designating particular tasks or operations as ones where protectors must be used.

Regulation 8

Maintenance and use of equipment

(1) The employer shall –

(a) ensure so far as is practicable that anything provided by him in compliance with his duties under these Regulations to or for the benefit of an employee, other than personal hearing protectors provided under regulation 7(1), is fully and properly used; and

(b) ensure that anything provided by him in compliance with his duties under these Regulations is maintained in an efficient state, in efficient working order and in good repair.

(2) Every employee shall –

(a) make full and proper use of personal hearing protectors provided to him by his employer in compliance with regulation 7(2) and of any other control measures provided by his employer in compliance with his duties under these Regulations; and

(b) if he discovers any defect in any personal hearing protectors or other control measures as specified in sub-paragraph (a) report it to his employer as soon as is practicable.

8

Use and maintenance of noise-control equipment

95　You must ensure, so far as is practicable, that any noise-control equipment you put in place is fully and properly used. For example, if a noise enclosure is provided with an access door, you will need to make sure that the equipment is not operated while the door is open. You should make sure that adequate instructions and supervision are in place to achieve this.

96　You must also make sure that noise-control equipment is maintained. You should carry out regular checks and introduce a system for reporting any defects or problems to someone with authority and responsibility for remedial action. You will need to put right any deficiencies promptly.

97　Your programme of maintenance should include:

(a) inspecting the noise-control equipment (such as silencers or enclosures) periodically to make sure it is kept in good condition;

(b) monitoring the equipment's effectiveness. Spot checks of the noise level at pre-selected locations will usually be adequate;

(c) reporting the results of these checks to someone with responsibility and authority for taking remedial action.

Use and maintenance of hearing protectors

98　People are often reluctant to use hearing protectors, and even where they start to use them they can easily get out of the habit. You therefore need to introduce a systematic programme to ensure they are used, taking into account the following:

(a) the firm's *safety policy*, which should include a clear commitment to using personal protection;

Guidance

8

(b) *signs and warning notices* to ensure awareness of where and when protectors should be used (see paragraph 92);

(c) *clear responsibilities*. You should identify who is responsible for the hearing protection programme and the distribution and maintenance of protectors;

(d) *information, instruction and training* for all employees on the risks and the action they should take (see paragraphs 115-118);

(e) *records* which should include details of the issue of hearing protectors, arrangements for ensuring users know where and how to use them, and any problems people encounter when using them;

(f) *monitoring* including spot checks to find out whether the hearing protectors are being used. You should keep a record and introduce a system to enable people to report deficiencies to a person with responsibility and authority for remedial action. Where an employee is not using hearing protection properly you should ask them why, and either resolve the difficulty or give and record a verbal warning. Where people persistently fail to use protectors properly you should follow normal disciplinary procedures.

99   You will need to arrange for someone to inspect reusable hearing protectors periodically and to repair or replace them if necessary. If your employees use disposable protectors, you should check that supplies are continuously available, and fill dispensers up regularly. Make sure that dirt does not get into the dispensers themselves and that employees are not inserting the protectors with dirty hands. You should introduce a system for employees to report any damaged, defective or lost protectors.

100   You should make proper provision for storage of reusable protectors, such as:

(a)   storage bags for earmuffs;

(b)   clean lockers where employees can keep them with other clothing.

101   You should also ensure that any special cleaning materials needed to clean their hearing protectors are available to users.

Employees' duties

102   Employees have a duty to comply with and use the measures you provide under the Noise Regulations, including:

(a)   using noise-control measures, such as exhaust silencers and machine enclosures, in accordance with your instructions;

(b)   wearing hearing protection in accordance with instructions provided when exposed at or above the upper exposure action values and at all times in areas marked as hearing protection zones;

(c)   taking care of hearing protectors and noise-control equipment they need to use;

(d)   reporting, in accordance with your procedures, any defect found in the hearing protectors or other protective measures or any difficulties in using them.

Guidance

8

103 In addition, under the Health and Safety at Work etc Act 1974, employees are required generally to co-operate with their employer to enable the employer to carry out legal duties.

Regulation 9

Health surveillance

Regulation

(1)   If the risk assessment indicates that there is a risk to the health of his employees who are, or are liable to be, exposed to noise, the employer shall ensure that such employees are placed under suitable health surveillance, which shall include testing of their hearing.

(2)   The employer shall ensure that a health record in respect of each of his employees who undergoes health surveillance in accordance with paragraph (1) is made and maintained and that the record or a copy thereof is kept available in a suitable form.

(3)   The employer shall –

(a)   on reasonable notice being given, allow an employee access to his personal health record; and

(b)   provide the enforcing authority with copies of such health records as it may require.

(4)   Where, as a result of health surveillance, an employee is found to have identifiable hearing damage the employer shall ensure that the employee is examined by a doctor and, if the doctor or any specialist to whom the doctor considers it necessary to refer the employee considers that the damage is likely to be the result of exposure to noise, the employer shall –

(a)   ensure that a suitably qualified person informs the employee accordingly;

(b)   review the risk assessment;

(c)   review any measure taken to comply with regulations 6, 7 and 8, taking into account any advice given by a doctor or occupational health professional, or by the enforcing authority;

(d)   consider assigning the employee to alternative work where there is no risk from further exposure to noise, taking into account any advice given by a doctor or occupational health professional; and

(e)   ensure continued health surveillance and provide for a review of the health of any other employee who has been similarly exposed.

(5)   An employee to whom this regulation applies shall, when required by his employer and at the cost of his employer, present himself during his working hours for such health surveillance procedures as may be required for the purposes of paragraph (1).

9

Guidance

9

104 Health surveillance is a programme of systematic health checks to identify early signs and symptoms of work-related ill health and to allow action to be taken to prevent its progression. It is also useful in monitoring the effectiveness of your controls, though it is not in itself a control measure or a substitute for controlling risk at source. Suitable health surveillance usually means regular hearing checks (audiometric testing). Part 6 provides information on health surveillance and sets out what you have to do. Some terms in regulation 9 are explained in paragraphs 105-114.

Guidance

9(1)

"If the risk assessment indicates that there is a risk to health"

105 Regulation 9(1) requires you to provide suitable health surveillance where the risk assessment indicates a risk to workers' health, ie a risk from exposure to noise without taking account of the noise reduction provided by hearing protection (see paragraph 31). The results of your health surveillance will enable you to check, among other things, whether your hearing protection programme has prevented hearing damage.

106 There is strong evidence to show that regular exposure above the upper exposure action values can pose a risk to health. You should therefore provide health surveillance to workers regularly exposed above the upper exposure action values. Where exposure is between the lower and upper exposure action values, or where employees are only occasionally exposed above the upper exposure action values, you should provide health surveillance if you find out that an individual may be particularly sensitive to noise. This may be from past medical history, audiometric test results from previous jobs, other independent assessments or a history of exposure to noise levels exceeding the upper exposure action values. A few individuals may also indicate a family history of becoming deaf early on in life.

107 If you provide health surveillance in accordance with paragraph 105 there should be no need for employees to seek separate advice from a doctor as allowed under Article 10.2 of the European Directive. Access to such services is available to employees through the National Health Service, and it is therefore not necessary to make specific provision for it under these Regulations.

Guidance

9(2) and (3)

"Health record"

108 These records will contain information on the outcome of the health surveillance and information on the individual's fitness to work in noisy environments. They should not contain confidential medical information, which should be kept in the medical record held by an occupational health professional. If your firm should cease to trade you should offer the health records to the individual concerned.

Guidance

9(4)

Action required when health surveillance reveals that an employee has suffered ill health as a result of exposure to noise

109 The doctor or occupational health professional who has made the diagnosis will explain the significance of the results to the employee and give advice on the risks of continuing to be exposed to noise at work.

110 The doctor or occupational health professional will inform you of the findings of the health surveillance procedures, in particular whether or not the employee is fit to continue work involving noise. However, they will not give medical-in-confidence information to you without the written consent of the employee.

111 You should prevent further harm to the individual by acting on advice from the doctor or occupational health professional and, where necessary, removing the employee from exposure to noise. You should review your risk assessment to decide whether to take action to protect the rest of the workforce. Where other workers are similarly exposed to noise you should arrange for their health to be reviewed.

112 By receiving an analysis of the anonymised health results of groups of employees, you can gain an insight into how well your noise-control and hearing-conservation programme is working. Such information should be suitably adapted to protect individuals' identities and be made available to safety or employee representatives.

Attendance for health surveillance

113 Regulation 9(5) requires your employees to co-operate with your health surveillance programme by attending their health surveillance appointments. However, you must arrange for this as part of their paid employment and cover any costs.

Consultation with employees and their representatives

114 You should consult with the employees concerned and their employee or safety representatives before introducing health surveillance. It is important that they understand that the aim of health surveillance is to protect them from developing advanced symptoms of ill health. You will need their understanding and co-operation if health surveillance is to be effective.

Regulation 10

Information, instruction and training

(1)   Where his employees are exposed to noise which is likely to be at or above a lower exposure action value, the employer shall provide those employees and their representatives with suitable and sufficient information, instruction and training.

(2)   Without prejudice to the generality of paragraph (1), the information, instruction and training provided under that paragraph shall include –

(a)   the nature of risks from exposure to noise;

(b)   the organisational and technical measures taken in order to comply with the requirements of regulation 6;

(c)   the exposure limit values and upper and lower exposure action values set out in regulation 4;

(d)   the significant findings of the risk assessment, including any measurements taken, with an explanation of those findings;

(e)   the availability and provision of personal hearing protectors under regulation 7 and their correct use in accordance with regulation 8(2);

(f)   why and how to detect and report signs of hearing damage;

(g)   the entitlement to health surveillance under regulation 9 and its purposes;

(h)   safe working practices to minimise exposure to noise; and

(i)   the collective results of any health surveillance undertaken in accordance with regulation 9 in a form calculated to prevent those results from being identified as relating to a particular person.

(3)   The information, instruction and training required by paragraph (1) shall be updated to take account of significant changes in the type of work carried out or the working methods used by the employer.

(4)   The employer shall ensure that any person, whether or not his employee, who carries out work in connection with the employer's duties under these Regulations has suitable and sufficient information, instruction and training.

Information, instruction and training for employees

115   It is important that employees understand the level of risk they may be exposed to, how it is caused and the possible effects and consequences. You should be as informative and open as you can to your exposed workers and to their employee and safety representatives. Regulation 10(2) lists some of the issues that must be covered, but it is not exhaustive.

116   It is important to tell employees:

(a)   the likely noise exposure and the risk to hearing the noise creates;

(b)   what you are doing to control risks and exposures;

(c)   where and how people can obtain hearing protectors;

(d)   how to report defects in hearing protectors and noise control equipment;

(e)   the employee's duties under the Noise Regulations;

(f)   what health surveillance employees will be provided with and how you are going to provide it;

(g)   what symptoms they should look out for (such as difficulty in understanding speech in conversation or when using the telephone, or permanent ringing in the ears), to whom they should report them and how they should report them.

117   You can provide the information, instruction and training in different ways, eg verbal explanations, computer-based training, videos, leaflets. The important thing is to make sure you give the information in a way in which the employee can understand it. You will need to reinforce the messages from time to time, and you should draw employees' attention to any relevant advice provided by HSE and provide them with the HSE pocket card.[2]

118   Programmes for controlling noise exposure are more likely to succeed when there is co-operation between yourself and your employees. The involvement of safety representatives and other employee representatives will be invaluable in promoting this co-operation. Working with trade-union-appointed safety representatives or other employee representatives can be a very useful means of communicating health and safety matters in your workplace. You are required by the Safety Representatives and Safety Committees Regulations 1977[6] and the Offshore Installations (Safety Representatives and Safety Committees) Regulations

Guidance

10

1989[7] to make certain information available to safety representatives appointed under the Regulations. The representatives are entitled to inspect your documents. These will normally include records of risk assessments covering the employees represented. You should make sure the representatives know how the information can be obtained and give them any necessary explanations of their meaning.

119 There is also a duty on employers to provide information to employee representatives elected under the Health and Safety (Consultation with Employees) Regulations 1996,[8] which apply to groups of workers who are not covered by a trade-union-appointed safety representative.

Guidance

10(4)

Information, instruction and training in connection with the employer's duties

120 Anyone who helps you comply with your duties under the Noise Regulations (eg by making noise measurements, determining exposures or planning for control of risk through changes to industrial processes or working practices) must be competent to undertake the task. Whether you employ a consultant or use members of your staff for these purposes you must satisfy yourself of their competence and provide them with any information on the work necessary for them to undertake the tasks.

121 Part 2 paragraphs 184-187 contain guidance on appropriate levels of knowledge and expertise for competent assessment and management of noise risks.

Regulation 11

Exemption certificates from hearing protection

Regulation

11

(1) Subject to paragraph (2), the Executive may, by a certificate in writing, exempt any person or class of persons from the provisions of regulation 6(4) and regulation 7(1) and (2) where because of the nature of the work the full and proper use of individual hearing protectors would be likely to cause greater risk to health or safety than not using such protectors, and any such exemption may be granted subject to conditions and to a limit of time and may be revoked by a certificate in writing at any time.

(2) The Executive shall not grant such an exemption unless –

(a) it consults the employers and the employees or their representatives concerned;

(b) it consults such other persons as it considers appropriate;

(c) the resulting risks are reduced to as low a level as is reasonably practicable; and

(d) the employees concerned are subject to increased health surveillance.

Guidance

11

122 HSE may grant an exemption from the requirements not to exceed the exposure limit values and to provide hearing protection as long as it is satisfied that the health and safety of people who are likely to be affected by the exemption will not be prejudiced as a result. HSE may grant an exemption subject to time limitation and conditions, and may revoke it. You would be required to agree with HSE a programme to make sure you control and check noise exposure, and introduce improvements as soon as reasonably practicable.

Guidance

123 Any exemption under this regulation and regulations 12 and 13 will not remove the duty under regulation 6(1) to eliminate the risk from noise or reduce it to as low a level as is reasonably practicable.

124 HSE may only consider exemptions where:

(a) the compulsory use of hearing protectors might increase danger overall, outweighing the risk of hearing damage; or

(b) it is not practicable to use hearing protectors meeting the requirements of regulation 7(4)(a), as long as people wear the most appropriate hearing protection.

Applications for exemptions

125 You should make any application for exemption to the authority responsible for enforcing health and safety legislation in your premises. If in doubt about who this is, consult your HSE local office. There is no standard form, but you will need to supply full supporting information prepared by someone with a good understanding of the problem and ways of combating it, including:

(a) the source of the noise that makes hearing protectors necessary;

(b) what the risk is if your employees use hearing protectors;

(c) what you are currently doing to protect against that risk;

(d) the existing arrangements for identifying individuals who might have particular difficulty in hearing warning sounds (eg because of hearing loss) and for ensuring their safety;

(e) how possible it is to reduce noise exposures in the short term and through a planned long-term noise-reduction programme;

(f) how possible it is to provide alternative safety arrangements (eg for warning employees or reducing the risk).

How HSE will deal with your application for an exemption

126 HSE will acknowledge your application promptly and give you the opportunity to discuss it if you want. The final decision on your application will be given in writing.

127 If HSE intends to vary any conditions in or revoke an exemption, usually after an exchange of views, you will be informed in writing.

11

Regulation 12

Exemption certificates for emergency services

Regulation

12

(1)   Subject to paragraph (2), the Executive may, by a certificate in writing, exempt any person or class of persons from the provisions of regulation 6(4) and regulation 7(1) to (3) in respect of activities carried out by emergency services which conflict with the requirements of any of those provisions, and any such exemption may be granted subject to conditions and to a limit of time and may be revoked by a certificate in writing at any time.

Regulation 12	*(2) The Executive shall not grant such an exemption unless it is satisfied that the health and safety of the employees concerned is ensured as far as possible in the light of the objectives of these Regulations.*
Guidance 12	128 Any emergency service wishing to seek exemption under this regulation should contact HSE for further advice. HSE is likely only to consider applications made in relation to an emergency service as a whole rather than from local units.

Regulation 13

Exemptions relating to the Ministry of Defence

Regulation 13

(1) Subject to paragraph (2), the Secretary of State for Defence may, by a certificate in writing, exempt any person or class of persons from the provisions of regulation 6(4) and regulation 7(1) to (3) in respect of activities carried out in the interests of national security which conflict with the requirements of any of those provisions, and any such exemption may be granted subject to conditions and to a limit of time and may be revoked by a certificate in writing at any time.

(2) The Secretary of State shall not grant such an exemption unless he is satisfied that the health and safety of the employees concerned is ensured as far as possible in the light of the objectives of these Regulations.

Regulation 14

Extension outside Great Britain

Regulation 14

These Regulations shall apply to and in relation to any activity outside Great Britain to which sections 1 to 59 and 80 to 82 of the 1974 Act apply by virtue of the Health and Safety at Work etc. Act 1974 (Application Outside Great Britain) Order 2001[(a)] *as those provisions apply within Great Britain.*

(a) SI 2001/2127.

Guidance 14

129 The Noise Regulations apply to all work activities on offshore installations, wells, pipelines and pipelines works and to certain connected activities within the territorial waters of Great Britain or in the designated areas of the UK Continental Shelf. The Noise Regulations also apply to certain other activities within territorial waters, including the construction and operation of wind farms.

Regulation 15

Revocations, amendments and savings

Regulation 15

(1) In –

(a) regulation 3(3)(e) of the Personal Protective Equipment at Work Regulations 1992;[(a)] *and*

(b) regulation 12(5)(d) of the Provision and Use of Work Equipment Regulations 1998,[(b)]

for the reference in each case to the Noise at Work Regulations 1989[(c)] *there shall be substituted a reference to these Regulations.*

(a) SI 1992/2966.
(b) SI 1998/2306.
(c) SI 1989/1790.

Regulation
15

(2)   The revocations listed in Schedule 3 are made with effect from the coming into force of these Regulations.

(3)   In respect of the music and entertainment sectors only, the amendment and revocations in paragraphs (1) and (2) shall not come into force until 6th April 2008 and the provisions covered by those paragraphs shall continue in force, where applicable, until that date.

Schedule 1

Part 1: Daily personal noise exposure levels

Regulation 2(1)

1 The daily personal noise exposure level, $L_{EP,d}$, which corresponds to $L_{EX,8h}$ defined in international standard ISO 1999: 1990 clause 3.6, is expressed in decibels and is ascertained using the formula:

$$L_{EP,d} = L_{Aeq,T_e} + 10 \log_{10} \left(\frac{T_e}{T_0} \right)$$

where –

T_e is the duration of the person's working day, in seconds;

T_0 is 28,800 seconds (8 hours); and

L_{Aeq,T_e} is the equivalent continuous A-weighted sound pressure level, as defined in ISO 1999: 1990 clause 3.5, in decibels, that represents the sound the person is exposed to during the working day.

2 If the work is such that the daily exposure consists of two or more periods with different sound levels, the daily personal noise exposure level ($L_{EP,d}$) for the combination of periods is ascertained using the formula:

$$L_{EP,d} = 10 \log_{10} \left[\frac{1}{T_0} \sum_{i=1}^{i=n} \left(T_i \, 10^{0.1 \left(L_{Aeq,T} \right)_i} \right) \right]$$

where –

n is the number of individual periods in the working day;

T_i is the duration of period i;

$(L_{Aeq,T})_i$ is the equivalent continuous A-weighted sound pressure level that represents the sound the person is exposed to during period i; and

$\sum_{i=1}^{i=n} T_i$ is equal to T_e, the duration of the person's working day, in seconds.

Part 2: Weekly personal noise exposure levels

Regulation 2(1)

The weekly personal noise exposure, $L_{EP,w}$, which corresponds to $\overline{L}_{EX,8h}$ defined in international standard ISO 1999: 1990 clause 3.6 (note 2) for a nominal week of five working days, is expressed in decibels and is ascertained using the formula:

$$L_{EP,w} = 10 \log_{10} \left[\frac{1}{5} \sum_{i=1}^{i=m} 10^{0.1 \left(L_{EP,d} \right)_i} \right]$$

where –

m is the number of working days on which the person is exposed to noise during a week; and

$(L_{EP,d})_i$ is the $L_{EP,d}$ for working day i.

1

Schedule 2

Schedule

2

Peak sound pressure level

Regulation 2(1)

Peak sound pressure level (L_{Cpeak}), is expressed in decibels and is ascertained using the formula:

$$L_{\mathrm{Cpeak}} = 20 \log_{10} \left[\frac{p_{\mathrm{Cpeak}}}{p_0} \right]$$

where –

p_{Cpeak} is the maximum value of the C-weighted sound pressure, in Pascals (Pa), to which a person is exposed during the working day; and

$p_0 = 20 \ \mu\mathrm{Pa}$.

Schedule 3

Schedule

3

Revocations

Regulation 15(2)

Regulations revoked	References	Extent of revocation
The Noise at Work Regulations 1989	S.I. 1989/1790	The whole Regulations
The Quarries Regulations 1999	S.I. 1999/2024	Schedule 5 Part II insofar as it amends regulation 2 of the Noise at Work Regulations 1989
The Offshore Electricity and Noise Regulations 1997	S.I. 1997/1993	Regulation 3
The Personal Protective Equipment at Work Regulations 1992	S.I. 1992/2966	Schedule 2 Part IX
The Health and Safety (Safety Signs and Signals) Regulations 1996	S.I. 1996/341	Schedule 3 Part II paragraph 1(a) and (b)
The Offshore Installations and Wells (Design and Construction, etc.) Regulations 1996	S.I. 1996/913	Schedule 1 paragraph 46(b) and the word "noise" in paragraph 59(b)

PART 2: MANAGING NOISE RISKS – ASSESSMENT AND PLANNING FOR CONTROL

Overview

- How do I assess the risks to health and safety from noise at work?

- How do I decide what to do to control the risks?

- How do I plan the programme of control measures?

130 Exposure to noise at work can produce risks to employees' health and safety. You must do all that is reasonably practicable to eliminate these risks, or reduce them to a minimum. To do this effectively requires a systematic approach to the management of noise risks, covering risk assessment, planning how to control risks and putting the plan into action.

Skills and knowledge

131 To carry out the tasks involved in managing noise risks requires competence (skills and knowledge) in particular areas. You may have people within your workforce who have some of the necessary competencies, or who could acquire them with some training. Where the skills and knowledge for particular tasks are not available in-house you should call in external assistance, such as consultants, to carry out the work. Part 2 outlines what skills and knowledge are required at each stage, so that you can decide on what best suits your needs and circumstances.

Assess risks due to noise

132 The process of assessing risks to health and safety due to noise exposure is in five stages:

Stage 1   Is there risk due to noise?

Stage 2   Who might be harmed and how?

Stage 3   Evaluate the risks and develop a plan to control them.

Stage 4   Record the findings.

Stage 5   Review the risk assessment.

Stage 1   Is there risk due to noise?

Skills and knowledge required

- Understand the work going on.

- Understand how risks can arise from noise exposure.

- Be able to identify potentially problematic noise sources.

- Be able to obtain and understand noise information from machinery suppliers.

133 Your first step in managing noise risks is to decide whether the noise to which your employees are exposed may lead to risks to their health and safety.

134 Identifying whether there are noise risks should not be treated as a complex task. You should be able to come to a decision quite quickly using what you know about your business and the work that your employees do or by making simple observations in the workplace. If you can answer 'yes' to any of questions in the 'Noise hazard checklist', you probably have noise risks which need managing.

Noise hazard checklist

■ **Do you work in a noisy industry**, eg construction, demolition or road repair; woodworking; plastics processing; engineering; textile manufacture; general fabrication; forging, pressing or stamping; paper or board making; canning or bottling plant; foundries?

■ Do your employees use **noisy powered tools or machinery** for more than **half an hour each day** in total?

■ Are there noises due to impacts (such as hammering, drop forging, pneumatic impact tools etc), explosive sources such as cartridge-operated tools or detonators, or guns?

■ Are there areas of the workplace where noise levels could interfere with warning or danger signals?

'Listening checks'

■ Are employees exposed to noise which makes it necessary to shout to talk to someone **1 m away**, for more than about **half an hour per day** in total? *The noise level here is probably 90 dB or more.*

■ Are employees exposed to noise which makes it necessary to shout to talk to someone **2 m away**, for more than about **two hours per day** in total? *The noise level here is probably 85 dB or more.*

■ Is conversation at 2 m possible, but **noise is intrusive** - comparable to a busy street, a typical vacuum cleaner or a crowded restaurant - for more than about **six hours per day** in total? *The noise level here is probably 80 dB or more.*

---

135 You could also use the information provided by machinery suppliers as an indication of whether there is likely to be a noise problem. Suppliers of machinery are legally required to provide information on the noise emissions from their machinery (see Appendix 4).

136 Any audible sound should be considered as noise and be part of a person's noise exposure. This includes speech, music, noise from communication devices or personal stereos, as well as the noise of machinery and work processes.

137 Your employees may be at risk from the noise created by people who are not your employees, eg if your employees visit other workplaces, or work in places where a number of different employers are carrying out work (eg on a construction site). In these cases you will need to exchange information with any other employers concerned to decide whether there are noise risks. Similarly, you should provide information to other employers who have employees who may be affected by the work you are doing.

138 If you decide that you do have noise risks that need to be managed, then you will need to go on to evaluate those risks and plan how you will control them. If you are in any doubt as to whether there are noise risks, it is advised that you assume there are, and proceed accordingly.

139 If you are satisfied that your employees are not at risk from noise, you should document this conclusion as part of your general risk assessment procedures. If circumstances change which may affect the noise exposure of your employees you will need to review this conclusion.

Stage 2 Who might be harmed and how?

Skills and knowledge required

■ Understand all work going on.

■ Understand and be able to identify potential for harm and factors which influence risk, including but not limited to the list in regulation 5(3).

140 You should identify which employees are likely to be affected by the noise and how. You will already have some idea of which employees are at risk, but you need to ensure that all employees at risk have been identified. For example, consider not just the person operating a noisy tool or machine, but also other people working nearby who may also be affected. Consider also people who move between different jobs or types of work during the day, and make sure you understand their patterns of noise exposure. Remember to include people who are not your employees but who may be affected by the work you do, eg visitors or subcontractors working on your site or alongside your employees.

141 In considering the potential for people to be harmed, you mainly need to think about hearing damage (deafness, tinnitus or other hearing problems). But you also need to consider risks to safety which can arise from working in a noisy environment, such as noise interfering with communications, warning signals and the ability to pick up audible signs of danger. These risks to safety may be more closely related to the level of noise at a particular point in time rather than to daily personal noise exposure.

142 You will need to take into account employees with pre-existing hearing conditions, those with a family history of deafness (if known), pregnant women and young people.

Stage 3 Evaluate the risks and develop a plan to control them

Skills and knowledge required

■ Be able to estimate noise exposure and make judgements on likely exposure.

■ Understand exposure action and limit values, and know what legal duties apply.

■ Obtain and understand good practice and industry standards for noise control.

■ Be able to prioritise controls and tackle immediate risks.

■ Recognise where specific skills are required, and be able to access further competent advice.

143 To evaluate the risks from noise you need to assess the noise exposure of your employees, in terms of daily or weekly personal noise exposure and exposure to peak noise. You then need to compare your assessment of exposure with the action and limit values set out in the Noise Regulations (regulation 4).

144 The results of this comparison, in combination with consideration of other relevant risk factors and information on the sources and circumstances of noise exposure, will determine your legal duties under the Noise Regulations, and allow you to evaluate whether risks from noise exposure are reduced to the lowest level reasonably practicable. You can then begin to develop an action plan to control risks from noise.

Assess exposure to noise

145 To assess a worker's daily personal noise exposure you need information on:

- the average noise level (L_{Aeq}) to which the worker is exposed during the tasks which make up the working day; and

- the length of time the worker spends on each of the tasks.

146 Where workers are regularly exposed to steady noise throughout the working day (eg in a weaving shed), or to intermittent but regular periods of steady noise (eg the operator of an automatic lathe), estimating exposure is relatively straightforward. For situations where exposures are irregular, where workers intermittently use a variety of different machines, or spend time in different areas, determining a typical or likely exposure can be more complex. It is advisable to adopt a worst-case approach in these situations.

147 You are not required to make a highly precise assessment of noise exposure. However, your estimate of exposure must be reliable, and precise enough for you to be able to assess whether any exposure action values are likely to be exceeded.

148 To demonstrate that your estimate is reliable, you must be able to show that you used data which is representative of your employees' exposure to noise under the specific circumstances of their exposure, taking account of their particular work practices, and that you took account of uncertainties. Uncertainties may arise from measurement and sampling techniques, how representative the data sources are, and variations in the work.

Determine the noise level

149 The average noise level (L_{Aeq}) may be derived from measurements made in your workplace. It may also be derived from other sources of data, such as published information on noise levels or information from machinery manufacturers and suppliers (see Part 4). The most important factor is how representative the data are of your work situation. If you use data that are not based on measurements in your workplace, you are likely to have to make more effort to demonstrate that the data are representative and apply a greater uncertainty factor. You may find it helpful to arrive at a first approximation of the noise levels from other sources of data, resorting to measurements where you find that you cannot reliably say whether any exposure action values are exceeded.

150 Advice on measuring noise exposure in the workplace is given in Appendix 1. Information on the evaluation of peak noise exposure is given in Appendix 2.

Determine the duration of exposure

151 The best way to determine how long people are exposed to levels of noise during their work is by direct observation of the work going on, and discussions with employees and their supervisors.

152 When determining an appropriate value for duration to estimate noise exposure, you need to take account of how the information on noise levels was obtained. This is particularly important if noise exposures are intermittent or cyclic during a particular task or job. For example, if the noise level information is an average relating to the whole job within which the intermittent or cyclic exposures occur, then the duration of exposure should be determined as the duration of that job during the working day. If, however, the noise level information is based on the level during the period that the noise is present, then the duration of exposure should be determined as the length of time that the intermittent or cyclic noise is present.

Determine daily noise exposure

153 Schedule 1 Part 1 to the Noise Regulations sets out the mathematical relationship between time-averaged noise level and daily exposure, and provides a formula for combining noise exposure from multiple tasks to calculate the daily personal noise exposure. In Schedule 1 Part 2 a formula for calculating weekly noise exposure is given. Electronic spreadsheets are available on the HSE website which do these calculations for you (www.hse.gov.uk/noise). Simple methods for determining daily and weekly personal noise exposure using 'ready-reckoners' are given in paragraphs 154-160.

Noise exposure ready-reckoner

154 Table 2 shows a ready-reckoner that provides a simple way of working out the daily personal noise exposure of employees, based on the level of noise and duration of exposure. It can be used for situations where the level of noise is steady throughout the day, or where noise exposure is variable throughout the day due to different jobs or the type of work being done. It provides a way of working out 'noise exposure points' for individual jobs that can be combined to give the total exposure points for a day, and so finding out the daily exposure. Additionally, noise exposure points can be used to prioritise the noise-control programme, by showing which jobs or tasks make the greatest contribution to the total noise exposure. Tackling these noise sources will have the greatest effect in reducing personal noise exposures.

155 Noise exposure points are similar to the fractional exposure values that have appeared in previous HSE guidance on noise at work. Noise exposure points have replaced fractional exposures because they more readily apply to the exposure action values in the Noise Regulations 2005. To convert from fractional exposure values to noise exposure points, multiply the fractional exposure by 320.

156 In the noise exposure points scheme, the upper exposure action value (an $L_{EP,d}$ of 85 dB) is 100 points, and the lower exposure action value (an $L_{EP,d}$ of 80 dB) is 32 points.

157 The left section of Table 2 shows how noise level and duration of exposure are combined to give noise exposure points. The right section is used to convert total exposure points to daily personal exposure. A worked example using the ready-reckoner is also shown in Table 3.

158 The values in the tables have been rounded. Differences introduced by rounding will not significantly affect the results of your estimate of noise exposure.

Sound pressure level, L_{Aeq} (dB)	Duration of exposure (hours)							
	$^1/_4$	$^1/_2$	1	2	4	8	10	12
105	320	625	1250					
100	100	200	400	800				
97	50	100	200	400	800			
95	32	65	125	250	500	1000		
94	25	50	100	200	400	800		
93	20	40	80	160	320	630		
92	16	32	65	125	250	500	625	
91	12	25	50	100	200	400	500	600
90	10	20	40	80	160	320	400	470
89	8	16	32	65	130	250	310	380
88	6	12	25	50	100	200	250	300
87	5	10	20	40	80	160	200	240
86	4	8	16	32	65	130	160	190
85		6	12	25	50	100	125	150
84		5	10	20	40	80	100	120
83		4	8	16	32	65	80	95
82			6	12	25	50	65	75
81			5	10	20	40	50	60
80			4	8	16	32	40	48
79				6	13	25	32	38
78				5	10	20	25	30
75					5	10	13	15

Total exposure points	Noise exposure $L_{EP,d}$ (dB)
3200	100
1600	97
1000	95
800	94
630	93
500	92
400	91
320	90
250	89
200	88
160	87
130	86
100	85
80	84
65	83
50	82
40	81
32	80
25	79
20	78
16	77

Table 2 Noise exposure ready-reckoner

Noise exposure points – Worked example

An employee has the following typical work pattern: five hours working where a 'listening check' (see 'Noise hazard checklist') suggests the noise level is around 80 dB; two hours at a machine for which the manufacturer has declared 86 dB at the operator position (a 'listening check' suggests this is about right); 45 minutes on a task where noise measurements have shown 95 dB to be typical.

Noise level	Duration	Notes	Exposure points
80	5 hrs	No column for 5 hours, so add together values from 4 and 1 hour columns in row corresponding to 80 dB.	16 + 4 = 20
86	2 hrs	Directly from table	32
95	45 minutes	No column for 45 minutes, so add together values from 30 and 15 minute columns in row corresponding to 95 dB.	65 + 32 = 97
		Total noise exposure points	149
		$L_{EP,d}$	86 to 87 dB

Sound pressure level, L_{Aeq} (dB)	Duration of exposure (hours)								Total exposure points	Noise exposure $L_{EP,d}$ (dB)
	1/4	1/2	1	2	4	8	10	12		
95	32	65	125	250	500	1000			800	94
94	25	50	100	200	400	800			630	93
93	20	40	80	160	320	630			500	92
92	16	32	65	125	250	500	625		400	91
91	12	25	50	100	200	400	500	600	320	90
90	10	20	40	80	160	320	400	470	250	89
89	8	16	32	65	130	250	310	380	200	88
88	6	12	25	50	100	200	250	300	160	87
87	5	10	20	40	80	160	200	240	130	86
86	4	8	16	32	65	130	160	190	100	85
85	3	6	12	25	50	100	125	150	80	84
84		5	10	20	40	80	100	120	65	83
83		4	8	16	32	65	80	95	50	82
82			6	12	25	50	65	75	40	81
81			5	10	20	40	50	60	32	80
80			4	8	16	32	40	48	25	79
79				6	13	25	32	38	20	78
78				5	10	20	25	30	16	77
75					5	10	13	15		

Table 3 Worked example of noise exposure ready-reckoner

This pattern of noise exposures gives an $L_{EP,d}$ of between 86 and 87 dB. The priority for noise control or risk reduction is the task involving exposure to 95 dB for 45 minutes, since this gives the highest individual noise exposure points.

Weekly noise exposure ready-reckoner

159 In the circumstances outlined in Part 1 (see paragraphs 27-29) the weekly noise exposure level rather than a daily exposure level can be used as an indicator of risk.

160 The weekly noise exposure level ($L_{EP,w}$) takes account of the daily personal noise exposures for the number of days worked in a week (up to a maximum of seven days). It may be calculated using the formula given in Schedule 1 Part 2 of the Regulations. A ready-reckoner for calculating weekly exposure from the daily exposures for up to seven days is given in Table 4. An electronic spreadsheet for calculating weekly exposure can be found on the HSE website (www.hse.gov.uk/noise).

Daily noise exposure, $L_{EP,d}$	Points								Total exposure points	Weekly noise exposure, $L_{EP,w}$
	Day 1	Day 2	Day 3	Day 4	Day 5	Day 6	Day 7			
95	1000	1000	1000	1000	1000	1000	1000		5000	95
94	800	800	800	800	800	800	800		4000	94
93	630	630	630	630	630	630	630		3200	93
92	500	500	500	500	500	500	500		2500	92
91	400	400	400	400	400	400	400		2000	91
90	320	320	320	320	320	320	320		1600	90
89	250	250	250	250	250	250	250		1300	89
88	200	200	200	200	200	200	200		1000	88
87	160	160	160	160	160	160	160		800	87
86	130	130	130	130	130	130	130		630	86
85	100	100	100	100	100	100	100		500	85
84	80	80	80	80	80	80	80		400	84
83	65	65	65	65	65	65	65		320	83
82	50	50	50	50	50	50	50		250	82
81	40	40	40	40	40	40	40		200	81
80	32	32	32	32	32	32	32		160	80
79	25	25	25	25	25	25	25		130	79
78	20	20	20	20	20	20	20		100	78

Table 4 Weekly exposure ready-reckoner

Compare exposure to the exposure action values and find what duties apply

161 You need to compare your estimates of exposure against the lower and upper exposure action values to determine what specific duties apply to you in respect of your employees. Specific duties under regulations 6 (control of noise), 7 (hearing protection) and 10 (information, instruction and training) apply where particular exposure action values are likely to be exceeded.

162 To decide the likelihood of exposure action values being exceeded you will need to take account of the uncertainties in your estimate of exposure. If you estimate exposure as being close to an exposure action value you should proceed as if the action value has been exceeded.

163 The likelihood of exposure action values being exceeded can depend on whether particular jobs or activities take place on a daily, weekly or less frequent basis. For example, certain tasks may produce very high noise exposures but may be carried out infrequently. You must consider whether, if you assessed the noise exposure on a particular day, that day was representative of a typical day for that employee. Where workers' tasks vary from day to day, you should compare exposure at least from typical and worst-case working days against the exposure action values, and evaluate risks taking account of the pattern of daily exposure.

Are the risks as low as is reasonably practicable?

164 If employees' noise exposure is below the lower exposure action values, the risk of hearing damage is likely to be very small. What you need to do in these cases is make sure that the risks remain at this low level. By understanding why the risks are low, you will be better able to make sure that they remain that way, and to know when changes in the workplace could lead to increased risks. Make a record of your current situation, and make sure you have proper systems of maintenance, supervision and management in place to keep on top of the situation. If there are simple measures that can be taken to reduce noise further, it is recommended that you carry them out.

Consider your general duties to control noise risks

165 At this stage in the process of evaluating risks you should have a good understanding of the level of risk, the circumstances under which the risks occur and the sources of risk. You will have identified what specific duties apply in relation to the exposure action values. But you should keep in mind the general duty under regulation 6(1) to reduce risk to the lowest level reasonably practicable.

166 This means you need to consider, in your evaluation of risks, alternative processes, equipment and/or working methods which would make the work quieter or mean people are exposed for shorter times. You should be aware of current good practice or the standard for noise control within your industry, considering whether such measures are applicable to your work and adopting them where it is reasonably practicable to do so (see paragraphs 169-171).

167 You should also consider whether suitable replacement machinery is available which emits lower levels of noise or would lead to lower levels of noise exposure for particular tasks where employees' noise exposure is influenced by noise emissions from tools and machinery which you supply for use at work. For example, if a more efficient tool allows work to be carried out in less time, it may lead to lower noise exposures without necessarily emitting less noise during use.

Implementing a policy on the purchase and hire of tools and machinery which includes consideration of noise would allow you to address these issues.

168 Also, in evaluating risks you should identify where maintenance of tools, machinery and equipment is important to ensure that noise levels do not increase over time. When considering maintenance systems, you should consider the need to minimise noise emissions for these items as part of your general duty to control noise risks.

Control of noise exposures – Good practice and industry standards

Consider the basic controls

169 There are many ways of controlling noise. Basic noise-control measures can often be effective in reducing the noise produced by a machine or process, or the noise exposure of employees. You will find advice on a range of noise-control measures and their field of application in Part 3. Many of these measures are likely to meet the test of 'reasonable practicability' – ie the costs (time, trouble, expense) arising from putting them in place will not grossly outweigh the benefits. You should consider implementing these basic measures.

Where solutions are known – the industry standard

170 For a lot of machines and processes, there are well-known noise-control solutions. HSE publishes free information on controlling noise in processes associated with engineering, woodworking, food and drink, construction, foundries and agriculture, as well as case study examples of noise control. See HSE's website www.hse.gov.uk to view free publications. There are also other sources of information, for example trade associations. Where such 'industry standard' noise-control measures could be applied or adapted to reduce noise exposures in your workplace, then you should put them in place, unless you can show that to do so would not be reasonably practicable.

171 Any good practices you put in place should be documented and properly planned, resourced and carried out. They should be subject to maintenance procedures and periodic review, as appropriate.

Consider the exposure limit values

172 When evaluating the risks to your employees from noise, you need to take account of the exposure limit values. These are limits set both in terms of daily or weekly personal noise exposure ($L_{EP,d}$ or $L_{EP,w}$ of 87 dB) and peak noise (L_{Cpeak} of 140 dB). In checking that you have not exceeded the exposure limit values you can take account of the protection provided by personal hearing protection. Appendix 3 gives general advice on how to estimate the protection provided by hearing protectors, and Appendix 2 contains specific advice in relation to peak noise.

173 If your risk assessment shows that any of your employees are exposed above either of the exposure limit values, you must take immediate action. If the employees are not provided with and wearing hearing protection, you should straight away provide them with hearing protection which will at least reduce their exposure to below the limit value, as long as personal hearing protection will be effective. If the exposure limit values are exceeded, even taking account of the hearing protection, then you should reduce exposure immediately, even if that means stopping the work.

174 Having dealt with the immediate problem of the limit value being exceeded, you should then identify the reasons why this happened, and put measures and systems in place to ensure that it does not happen again. You will need to review any noise-control and risk-reduction measures you already have in place, the suitability of any hearing protection you have supplied, and the systems for ensuring that hearing protection and any noise-control measures are properly used and maintained.

175 The exposure limit values take account of the effect of hearing protection, but they should not be used as a target for the performance of hearing protection. The Noise Regulations require that where hearing protectors are used they should eliminate the risks, or reduce risks to as low a level as is reasonably practicable. Part 5 gives guidance on selecting appropriate hearing protection.

Develop an action plan

176 Your noise action plan provides the vital link between the risk assessment and control of the risks identified. It is an important output of the noise risk assessment, and can be thought of as a statement of intent. It should set out a prioritised plan for investigating and introducing noise-control and risk-reduction measures, and any other measures required by the Noise Regulations. It should also list what you have done to control risks during the assessment process, eg actions you took to tackle the immediate risk, or those basic noise-control measures which you were able to put in place immediately. It is important to record not only what you have done, but also what you intend to do, so that all measures can be included in any subsequent review. The information you have gathered to carry out your risk assessment will help you to make your decisions about the action plan.

177 The action plan should contain the following:

■ A list of what you have done to tackle the immediate risks (eg providing appropriately selected hearing protection as a temporary measure where the upper exposure action values are exceeded, or taking action if the exposure limit values are exceeded).

■ The actions you are considering regarding your general duty to reduce risks, including:

❑ investigating the applicability of basic noise-control measures and relevant industry standards in noise control;

❑ implementing a positive hire and purchase policy;

❑ maintenance systems necessary to ensure minimum noise emissions from plant etc.

■ Your plans to develop a programme of noise-reduction measures, where exposure to noise exceeds the upper exposure action values, including how you will explore and prioritise options for controlling noise exposure control.

■ How you will provide suitable hearing protection, and set up hearing protection zones.

■ Arrangements for providing information, instruction and training for employees, including training on noise hazards and information on the measures that you have or will put in place to minimise risks.

■ Arrangements for providing health surveillance.

- Realistic time-scales for the work to be carried out.

- Assignment of tasks to named people or post holders within the company to be responsible for the various tasks.

- Assignment of a named person or post holder to be responsible, overall, for making sure that the plan is competently carried out.

178 The people responsible for the plan and the tasks within it should have enough authority, control and influence within the company to enable them to carry out their part of the plan properly. They may also need access to competent advice or competent services, and this need should be highlighted in the action plan.

179 Your priorities for action will inevitably be influenced by cost. Some noise-control measures can be expensive, and you may wish to assign these to a medium-term priority, depending on the degree of risk and protection available for employees in the meantime. However, there are many low- or no-cost control measures that could be implemented immediately.

Stage 4 Record the findings

Skills and knowledge required

- Understand what information to record, how and why to record it.

180 You should record the major findings of your risk assessment, and your action plan. The major findings are your estimates of daily personal noise exposure and peak noise exposure, your analysis of the risks and whether they are as low as reasonably practicable, any other information on which you based the evaluation of risks and the decisions on actions required under the Noise Regulations. The record will provide some evidence for the decisions that you made to comply with the law. A minimum adequate record will include details of:

- the workplaces, areas, jobs or people included in the assessment, including a description of the work going on;

- the date(s) that the assessment was made;

- the daily personal noise exposures of the employees or groups of employees concerned;

- the peak noise exposure levels of the employees or groups of employees concerned;

- the information used to determine noise exposure;

- if noise measurements have been made, relevant details of the measurements, including the person(s) responsible for carrying them out;

- any further information used to evaluate risks;

- the name of the person(s) responsible for making the risk assessment;

- your action plan to control noise risks.

Stage 5 Review the risk assessment

Skills and knowledge required

■ Know in what circumstances to review the assessment.

181 Your risk assessment should be reviewed if:

■ there is any reason to think that it does not reflect the current noise risk in your workplace. For example, if you change the way you work or the processes you use, bring in new machinery, stop using older machinery or alter shift patterns, the noise exposures of your employees are likely to change;

■ you become aware (eg through trade journals, industry groups or HSE publications) of new ways of working or improved noise-control techniques that could be applied to your workplace;

■ you have introduced noise-control measures following a previous assessment and need to determine their impact on employees' exposure;

■ health surveillance shows that employees' hearing is being damaged, suggesting that noise risks are not being properly controlled;

■ control measures that could not be justified when you originally conducted your risk assessment (probably on the grounds of costs) become reasonably practicable, eg because of changes in technology and cost.

182 A reassessment will often be simpler or involve less work than the original risk assessment. It is sensible to integrate it with other health and safety management activities, so that it is part of an ongoing programme which should pick up changes as they occur.

183 Even if you consider there have been no changes you should check every two years whether there is a need to review your risk assessment.

Competent advice and services

184 This book tells you at various stages the necessary skills and knowledge required to assess and manage noise risks. Where you do not have those skills and knowledge, or access to them from people within your workforce, then you will need to go outside the company for advice. In the longer term you may wish to organise some training for yourself or your employees in these areas, so that in the future you would have access to the necessary skills and knowledge in-house.

185 Whether carrying out the work yourself, appointing other people, or a combination of the two, you need to make sure that the assessment and management of noise risks is carried out in a competent manner based on competent advice. This means making sure that all people involved in the process have the necessary training and experience to carry out their part of the work. The ability to understand and apply this guidance may be more important than formal qualifications. But there are some areas (such as noise-control engineering) where the person providing the advice would be expected to have formal qualifications.

Training courses

186 A variety of short training courses is available organised both on a national basis and to meet local requirements. For example, courses organised on a national basis through technical institutions include sessions designed to provide training for the purposes of the Noise Regulations. Training on noise is also available in modules forming part of more general courses.

187 To provide cost-effective advice on engineering control, the person would need to make a more thorough study of the principles of noise-control engineering. Some universities and technical institutes provide specialised courses at a range of levels, and general courses on noise control engineering are available (see also paragraphs 238-239).

PART 3: PRACTICAL NOISE CONTROL

Overview

■ What are the practical steps to control noise?

■ What are the main noise-control techniques?

■ How do I avoid common problems?

■ Where might I need further advice?

188 Part 3 gives practical advice for employers on the options available for controlling noise. Basic techniques and methods for controlling noise are described, with examples. It is not intended as an exhaustive treatment on noise control. More detailed and specific guidance for particular industries and machines can be found in other HSE guidance (www.hse.gov.uk).

189 Some noise-control techniques are simple to put in place. Others require specialist skills and knowledge. There is the potential to misdirect effort, time and money in applying the wrong type of noise-control technique and therefore consultation with specialist noise-control engineers is often required.

190 The Noise Regulations give priority to the control of noise by technical or organisational means, as opposed to providing personal hearing protection because:

■ noise control is usually the most cost-effective solution for the longer term;

■ control of the noise risk at source protects a greater number of people in the surrounding working environment;

■ personal hearing protectors protect only the individual wearer and do not always give the protection expected.

191 You can make a positive contribution to noise control by making arrangements for effective controls. However, you should be aware of your limitations and understand when you need to take further advice. Noise control is not necessarily difficult or expensive. Effective, simple controls are available that can be carried out 'in-house'.

192 You should consider the methods of controlling noise in the following order:

■ Can risks be eliminated by doing the work in a different way, eliminating or minimising exposure to noise?

■ Can the work, process or machine be modified to reduce noise emissions?

■ Is it possible to replace the tools and equipment used with lower noise alternatives?

■ Can the workplace and workflow be arranged to separate people from the noise?

■ Can the noise reaching people be reduced by controlling it on its path from the source?

A noise-control checklist

Find out what and where the problem is and who is affected

Your risk assessment will have established who is exposed to noise and it will have highlighted tools, machines and processes that cause the noise exposure. With a little further investigation you can often identify the part of a machine that is responsible for producing the noise. This can be done by:

- observing and listening to the machine in question;

- identifying if parts of the machine are vibrating, which may be the source of the noise;

- treating the most dominant noise source first (there is little point in treating the easiest or cheapest first if it is not the main source of the noise);

- targeting control in areas where most employees work.

Consider the source of noise

- Is it cost effective to replace the machine by one with lower noise emissions?

- Could the machine be moved to an area with fewer employees without disrupting production?

- Is the machine being properly maintained?

- Is it possible to modify parts of the machine, eg by replacing components with ones designed to operate more quietly without affecting the safe operation of the machine?

Consider how the noise source radiates noise

- Are the machine's panels vibrating? Isolate the panels or add damping materials to them.

- Is the machine vibration entering the structure of the building (walls or floor)? Isolate the machine from the building with isolation mounts or isolated foundations.

- Is the noise caused by impacts from falling material? Add damping material to receiving trays and chutes or reduce the distance the material falls.

- Are solid guards attached to the machine around noisy components? Line the guards with sound-absorbing material and where possible seal gaps taking account of the need for ventilation.

- Is the major noise source caused by either the inlet or exhaust of air or gas from the machine? Fit an appropriate silencer to the inlet, exhaust or both.

- Is the noise caused by the sudden release of air from a compressed air system? Fit silencers or feed the exhaust away from the working area.

Consider the path of the noise

- Could you position the worker away from the source of noise? Doubling the distance can reduce the effect of the noise by 3 to 6 dB.

- Could you fit a suitably designed enclosure around a machine that does not require 'hands on' operation?

- Could you acoustically treat openings in the machinery or its enclosure into which material is placed or from which the product is removed?

- Could you fit acoustic ducts or quiet fans in enclosures where there may be a build-up of heat?

- Could you build a noise haven for employees supervising the operation of large machines where enclosing the whole machine would be difficult?

- Could you erect barriers or screens between different elements in the production process, separating quiet operations from noisy ones?

- Can absorptive materials be added to the building to reduce the reverberant noise (echoes)?

- Is the noise level constant and made up of low-frequency tones (eg, from fans and dryers)? If so, consider using active noise control.

Remember to check that noise levels have improved after the noise control has been fitted

193 Noise control can be divided into a number of basic methods, including organisational controls, workplace design, buying low-noise machinery, machine design, providing enclosures, screens, barriers and refuges, damping, isolation and fitting silencers.

Organisational controls

Plan and organise the work to reduce noise exposures

194 The way you plan, organise and lay out your work processes can have an effect on the noise exposure of individual employees, as can the tools and machines you use.

Job design

195 Noisy devices should only be used when they are actually needed. For example, the pneumatic ejector on a power press need be on only for the short time required to eject the product; the air supply should be 'pulsed' to operate only when the product needs removing.

Job rotation

196 Where some employees do noisy jobs all day, and others do quieter ones, consider introducing job rotation. This might need you to train employees to carry out other jobs. You should be aware that this system will reduce the noise exposure of some employees while increasing that of others, so care and judgement is needed. In addition, employees will need to be rotated away from

noisy jobs for a significant proportion of time to make an appreciable difference to their daily exposure. This is because daily exposure is dominated by time in noisy areas.

Different ways of working

197 Changes in technology can alter the machine or process, resulting in a lower noise exposure to the workforce. Sometimes a different way of working might avoid the need for a noisy operation. Examples of quieter processes, machines and activities include:

Change of process

- Use break-stem rivets instead of hot rivets set by traditional noisy hammering.

- Use welded or bolted construction instead of riveted construction in large-scale fabrications.

Change of machine

- Use hydraulic pressing of bearings into a casting instead of being driven in by hammering.

- Replace manual turning lathes on repetitive production with computer-controlled automatic machines which often have guards that offer some noise reduction and mean the operator needs to spend less time close to the source of the noise.

- Replace noisy compressed-air tools with hydraulic alternatives.

Change of activity

- Improve the quality of manufacturing to avoid later rework with potentially noisy processes, eg more accurate cutting of steel plate may eliminate noisy reworking with grinders or air chisels.

- Change the design of construction to avoid noisy processes (eg use retarding compounds to avoid scabbling concrete).

Workplace design

Design and lay out the workplace for reduced noise exposure

198 When considering a new workplace or modifying an existing one, noise emissions and noise exposure can be limited by careful choice of design, layout and the construction materials used for the building. For example, the appropriate use of absorption materials within the building can reduce or limit the effects of reflected sound (specialist help will be needed to put this into effect).

199 Noise risk management is a lot easier if you limit the number of employees exposed. Careful planning could segregate noisy machines from other areas where quiet operations are carried out, reducing the need for noise control after the workplace is in operation (see also the section on screens and barriers). The number of employees working in noisy areas should be kept to a minimum.

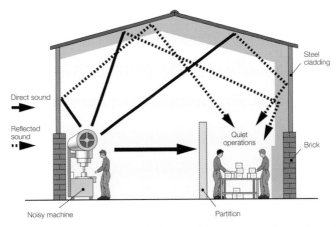

Figure 2 Noise paths found in a workplace. The quiet area is subjected to reflected noise from a machine somewhere else in the building

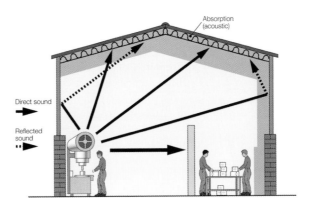

Figure 3 The correct use of absorption in the roof will reduce the reflected noise reaching the quiet area

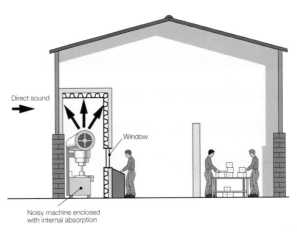

Figure 4 Segregation of the noisy operation will benefit the whole workplace

200 When considering using noise-absorbing materials to change the acoustic characteristics of a work area remember:

- environmental and workplace factors: absorption materials are available in forms which are designed to withstand physical impacts, and can be adapted to hygienic environments or where absorption of oil, water etc may be a problem;

- there may be a reduction in the natural light if absorption is placed in the roof;

- adding absorbent materials to walls and ceiling areas will only affect the reflected, reverberant sound – not the direct path of sound.

Low-noise machines

201 Selection of low-noise tools and machinery through a positive purchasing and hire policy can avoid the need to apply retrofit noise control. This could be the single most cost-effective, long-term measure you can take to reduce noise at work. Part 4 gives advice on selecting low-noise tools and machinery.

202 Your positive noise-reduction purchasing policy could involve:

- preparing a machine specification. Draw your suppliers' attention to the requirements of the Supply of Machinery (Safety) Regulations 1992[9] (see Part 4). Introduce your own company noise limit, ie a realistic low-noise emission level that you are prepared to accept from incoming plant and equipment given your circumstances and planned machine use;

- comparing the noise information declared by the manufacturer to identify low-noise machines;

- requiring a statement from all companies who are tendering or supplying, saying if their machinery will meet your company noise limit specification;

- discussing noise issues with the supplier of the machine. This may influence the design of future low-noise machines;

- where it is necessary to purchase noisy machinery, keeping a record of the reasons for decisions made to help with the preparation of future machine specifications with information on where improvements are necessary;

- using an agreed format for the presentation of results by suppliers;

- discussing your machinery needs and noise emission levels with your safety or employee representatives(s).

Machine design

Change the total or partial design of a machine, component or process

203 Machines and processes can be re-designed to generate less noise. This is something you could consider for existing machinery; new machinery should already be designed to produce as little noise as possible (see Part 4). Changes to the design of machines are likely to require some specialist advice from noise control engineers. Some particular examples of design changes are detailed in paragraphs 204-208.

Air turbulence noise

204 When any rotating part such as a fan blade or a woodworking cutter block passes close to a stationary part of the machine, noise is produced. If the distance between the rotating part and the stationary part is increased, the noise level will be reduced. Also if cutter blocks are fitted which have helical blades, the smooth transition of the curved cutting edge next to the stationary table instead of the abrupt impact of a normal blade will reduce the noise considerably.

205 **CAUTION:** Gaps between stationary and rotating parts of machinery are dangerous. You should not alter gaps without ensuring that the machinery can be used without risks to safety.

206 When air flows past an object or over sharp edges, turbulence is caused which produces noise. Also when air flows over cavities or voids a noise tone can be produced (similar to blowing over a milk bottle). Making edges as smooth as possible and removing voids or rounding the edges can reduce the level of noise created. Similarly, air flowing smoothly through ducts and pipes will produce less noise.

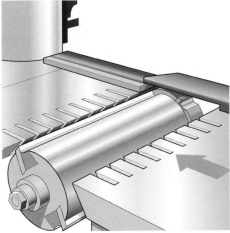

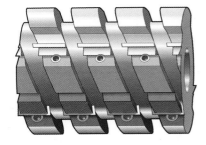

Figure 5 Slotted table lips on a planer reduce air turbulence and noise

Figure 6 Reduced-noise cutter block

Avoiding impacts

207 Noise generated by impacts, including components falling into chutes, bins and hoppers, and impacts generated by tooling can be considerable. Try to reduce the speed and/or height of falling objects and avoid impacts, or make arrangements to cushion them, eg:

- fit buffers on stops and rubber or plastic surface coatings on chutes, to avoid metal-to-metal impacts;

- apply a progressive cutting edge to punch tooling on power presses to reduce the impact noise;

- use conveyor systems, designed to prevent the components being transported from impacting against each other.

208 Try to limit or reduce the 'drop heights' of components. For example, components which are produced by pressings and are ejected and then dropped into a collecting bin can cause high noise levels. Reducing the force of the impact can reduce the noise levels.

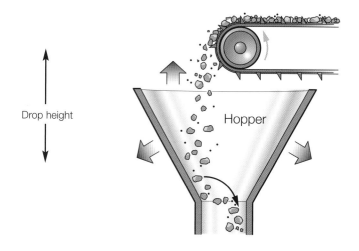

Drop height

Hopper

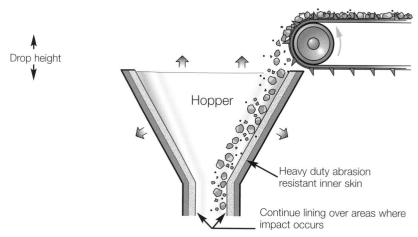

Drop height

Hopper

Heavy duty abrasion
resistant inner skin

Continue lining over areas where
impact occurs

Figure 7 Reduced drop heights and cushioned impacts reduce noise

Figure 8 Lining of a bowl feeder with rubber for reduced impact noise

Enclosure

Placing a sound-proof cover over the source of noise

209 Noisy machines can be enclosed fully, or a partial enclosure or an acoustic cover can be placed around a noisy part of a machine. Figure 9 outlines the features required of a typical machine enclosure.

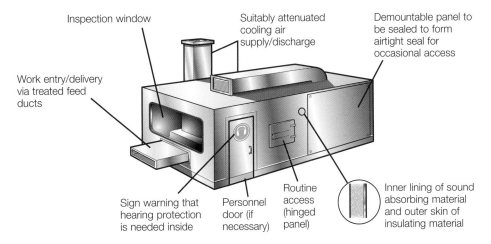

Figure 9 Features required of a typical machine enclosure

210 An efficient noise enclosure will provide:

■ a good quality dense insulating barrier to stop the noise from escaping (steel, brick etc);

■ sound-absorbing material on the inside to reduce the reflections and therefore reduce the build-up of noise in the enclosure;

■ double-glazed viewing windows;

■ good seals around openings – small leaks can dramatically reduce the effectiveness of the enclosure;

■ self-closing devices on any doors;

■ absorbent-lined cooling ducts;

■ absorbent-lined inlets and outlets for materials and services.

211 **CAUTION:** Enclosing machinery is likely to increase the temperature of the air inside the enclosure. Always provide adequate ventilation and cooling.

Screens and barriers

Placing an obstacle between the noise source and the people

212 Screens, barriers or walls can be placed between the source of the noise and the people to stop or reduce the direct sound. Barriers should be constructed from a dense material, eg brick or sheet steel, although chipboard and plasterboard can be used.

Figure 10 Moveable acoustic screen in an engineering workshop

213 Screens and barriers work best when they are placed close to the noise source or close to the people you are trying to protect. The higher and wider they are, the more effective they are likely to be. They work best in rooms with either high or sound-absorbent ceilings.

214 Covering the barrier or screen with noise-absorbing material on the side facing the noise source will have the added advantage of reducing the sound reflected back into that area containing the noise source. Those workplaces which have already been treated with sound-absorbing material will help to create conditions which will allow the screen or barrier to perform to its maximum potential, since in these cases the direct noise is likely to be the dominant source.

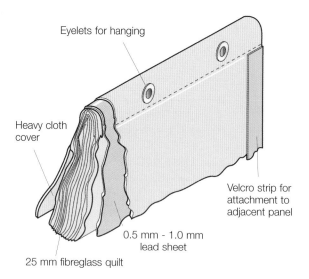

Eyelets for hanging

Heavy cloth cover

Velcro strip for attachment to adjacent panel

0.5 mm - 1.0 mm lead sheet

25 mm fibreglass quilt

Figure 11 Example of construction of a hanging, flexible acoustic barrier

215 **CAUTION:** Be aware of the following when using screens or barriers:

■   Screens and barriers may not work well for low frequencies.

■ They are best at reducing the direct noise, and may not affect reflected noise.

■ Always place the screen or barrier as close to the noise source or employee position as possible.

■ The screen or barrier should be made of a dense material, and should be lined with absorptive material facing the noise source.

■ Always consider other health and safety risks, such as safe movement of people and vehicles, when placing barriers in the workplace.

Refuges

Noise-reduced enclosures for people

216 Noise refuges can be a practical solution in situations where noise control is very difficult, or where only occasional attendance in noisy areas is necessary. The design of refuges will be similar to that of acoustic enclosures, although since the purpose is to keep noise out rather than in, lining the inner surfaces with acoustic absorbent material will not be necessary.

217 If machine controls are brought into the refuge, and thought is given to allowing remote monitoring or viewing of machinery and processes, it should be possible to minimise the amount of time that workers have to spend outside the refuge – so maximising the benefit of having the refuge. For example, a refuge that is used for only half a shift will achieve no more than 3 dB reduction in noise exposure.

218 Refuges must be acceptable to employees. This means they must be of a reasonable size, well lit and ventilated and have good ergonomic seating.

Figure 12 A noise refuge and control room

219 **CAUTION:** Check your refuge design for:

■ adequate ventilation;

■ good door and window seals;

■ self-closing doors;

■ dense construction materials, with plenty of acoustically double-glazed windows;

■ isolation from the floor to reduce structure vibrations;

■ size – is it large enough?

Damping

Adding material to reduce vibration and noise

220 Damping involves adding material to reduce vibrations and the tendency of machine parts to 'ring'.

221 Vibrations in a machine may affect objects that are connected to it, which in turn vibrate and give out noise. An example of this is where light steel panels are used to box in dangerous moving parts. These panels are excited by vibrations in the machine, which in turn cause noise to be radiated. Assess the flexibility of the panel and judge whether adding damping materials to the vibrating panels might reduce the noise.

222 Different ways of applying damping include:

■ applying treatments to sheet metal, such as spray-on or magnetic surface coatings or bonding two sheets together (eg a sheet of rubber bonded to a sheet of steel);

■ using materials such as sound-deadened steel with high damping capacity in the construction of machine casings;

■ attaching damping plates with bolts or spot welds to increase friction damping;

■ using secondary sheets of material (or lamination) to provide a degree of damping;

■ adding stiffness in the form of strengthening ribs to panels;

■ buying circular saw blades with a 'sandwich' damping layer, or having slots professionally laser-cut into blades to reduce the vibrations.

Figure 13 Damping material applied to transport chutes in the food industry

Figure 14 Damping compound being applied to metal decking

223 **CAUTION:** Damping will only be effective if the following points are observed:

■ Check at regular intervals that the damping material is not becoming detached from the machine.

■ Check if the damping material is deteriorating? If so, replace it.

■ Damping may only be effective for a limited range of noise frequencies.

■ Check you are using enough damping material.

Isolation

Separate the machine from its surroundings and supporting structures

224 Flexible isolation made of rubber or springs can be used to reduce the spread of structure-borne sound through a machine frame, eg:

■ isolate the bearings from a gearbox case to reduce the transmission of gear noise;

■ mount machines on the correct anti-vibration mounts to reduce the transmission of vibration into the structure of the workplace which may be radiated as noise;

■ fit anti-vibration mountings to reduce the transmission of sound from hydraulic power supply pipes to the cab floor on an earth-moving machine.

225 Anti-vibration mountings under machines are not usually very effective for reducing noise close to a machine, unless the floor on which it stands is unusually flexible, eg mezzanine floors, but they can be effective in reducing structure-borne noise causing nuisance in nearby areas/rooms and workstations.

Figure 15 Anti-vibration mounts isolate a grinder from the supporting surface

Figure 16 Detail of anti-vibration mount

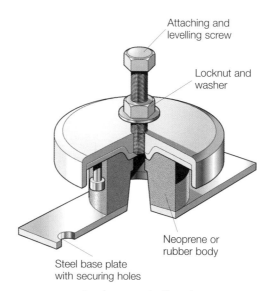

Figure 17 Rubber or neoprene in shear anti-vibration mount

226 **CAUTION**: Anti-vibration mounting, if not carried out correctly, can increase noise levels and affect the stability of machinery – seek specialist advice. Any supplier of anti-vibration mounts will provide specialist advice as part of the service.

Silencers

227 Silencers are attachments fitted to air or gas stream inlets and exhausts and can be used in many situations. Here are some examples:

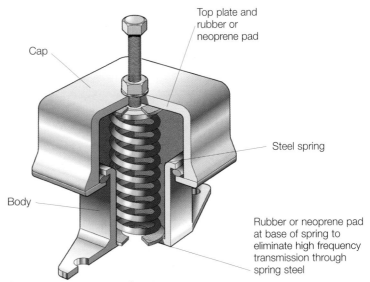

Figure 18 High-deflection spring anti-vibration mount

In pipes and ducts

228 Mufflers or silencers can reduce noise transmitted along pipes and ducts, eg:

■ exhaust and intake silencers on internal combustion engines;

■ duct silencers to control noise from the inlet and exhaust of ventilation fans.

Figure 19 A typical duct silencer

Air exhausts and jets

229 Silencers can reduce noise generated by turbulence at air exhausts and jets, eg:

■ a porous silencer for the exhaust of a pneumatic cylinder;

■ a silencer for the air supply to a shot-blaster's helmet;

■ use low-noise air nozzles (see Figure 20) and pneumatic ejectors and cleaning guns constructed on good aerodynamic principles, or substitute an alternative method of doing the job.

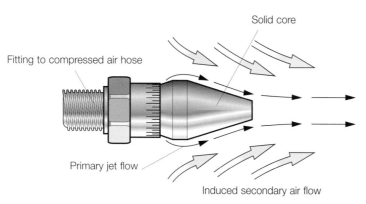

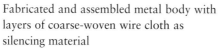

Figure 20 A low-noise (induced-flow) air nozzle

Figure 21 Multiple air exhausts piped to a manifold and silencer

Fabricated and assembled metal body with layers of coarse-woven wire cloth as silencing material

Machined metal body and thread with compacted wire wool as silencing material

Injection moulded plastic body and thread with sintered plastic or metal membrane

Figure 22 Typical air exhaust silencers

230 You should ensure that the quantity and pressure of compressed air supplied to equipment is matched to its needs. This can be achieved by providing each item of equipment with its own pressure-reducing valve. The supplies can be individually adjusted for a good compromise between reliable operation and noise. This has an added benefit of reducing the costs of the supply of compressed air.

231 **CAUTION:** Points to note when considering silencers:

■ Absorptive silencers are usually only good for medium to high noise frequencies.

■ Passive silencers are designed to reduce noise at a particular frequency, or over a small range of frequencies.

■ Contamination by debris (dust or liquids) will reduce the efficiency with time. Regular maintenance will maintain efficiency.

Active noise control

An electronically-controlled noise-reduction method

232 Active noise control is an electronically-controlled noise-reduction method and involves the reduction or cancellation of one sound by the introduction of a second 'opposite' sound.

233 The technique is most effective in reducing low-frequency noise. It has been used to control noise in ducted systems such as diesel engine exhausts and the low-frequency rumble from gas turbine stacks. It has also been used to extend the performance of hearing protection and noise-reducing helmets.

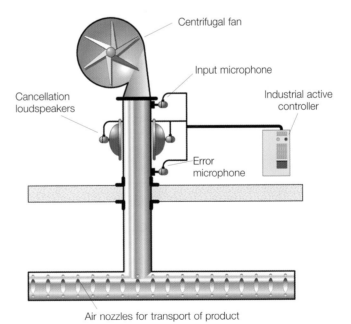

Figure 23 Active noise control in a pneumatic transport system

Distance

Increase the distance between the source of the noise and the people

234 Increasing the distance between a person and the noise source can reduce noise exposure considerably. Some examples of this are:

- direct the discharge from exhausts well away from workers, eg by fitting a flexible hose to discharge exhaust several metres away from the operator. Similarly, on a mobile machine powered by an internal combustion engine the exhaust can be kept well away from the driving position;

- use remote control or automated equipment to avoid the need for workers to spend long periods near to machines;

- separate noisy processes to restrict the number of people exposed to high levels of noise, eg test engines in test cells which need to be entered only occasionally, make arrangements for quiet inspection tasks to be carried out away from noisy manufacturing areas, and locate unattended air compressors and refrigeration plant in separate rooms.

Maintenance

Planned checking and maintenance of machinery

235 Machine maintenance can be critical in reducing noise. Machines deteriorate with age and use, and if not maintained are likely to produce more noise due to factors such as worn parts, poor lubrication and loose panels vibrating. Maintenance can, if carried out periodically, limit the increased noise emission due to wear.

236 Check with the operator that the noise level has not increased over time. If it has it is an indication that the machine requires attention. Have a reporting system in place so the operator can inform of problems.

237 Check that the noise-control features fitted to the machine have not deteriorated or been removed.

Specialist skills and training in noise control

238 While many noise-control techniques are simple to put in place and effective, other techniques require specialist skills and knowledge. Therefore if you are in any doubt as to whether you can identify for yourself the best means of controlling noise you are advised to consult a specialist noise-control engineer. In the longer term you should consider whether it would be beneficial to develop your own in-house skills and knowledge by providing training for yourself or some of your employees in noise control. Some technical colleges offer training in noise control.

239 Designing and putting in place effective noise control involves understanding the operational requirements of the work going on. A working knowledge of the process involved and alternative ways of doing the job are important factors. Particularly during development work it is likely to be necessary to identify and overcome problems encountered when introducing new ways of controlling the noise.

PART 4: SELECTING QUIETER TOOLS AND MACHINERY

Overview

■ How do I select low-noise tools and machinery?

■ What information should the manufacturer, supplier or hirer provide?

■ What are the limitations of the information?

240 Parts 1 to 3 identified tool and machine selection as an important step in the control of workplace noise exposure. Part 4 gives guidance for the employer on how to consider noise when selecting the tools and machinery that employees use at work. It sets out what the user (ie the employer who provides tools and machinery for employees) should expect from those supplying machinery (manufacturers, suppliers, importers and hire companies), as a result of the legal duties on manufacturers and suppliers under the Supply of Machinery (Safety) Regulations 1992, as amended (the Supply Regulations) and the Noise Emission in the Environment by Equipment for Use Outdoors Regulations 2001 (the Outdoor Equipment Regulations).[10] It also indicates special precautions that may be required in interpreting the information provided by manufacturers and suppliers.

Providing quieter machinery for your employees

241 You have a duty under the Provision and Use of Work Equipment Regulations 1998 (PUWER)[11] to ensure that tools and machinery that you provide for your employees are suitable for the work being done. This means you must take account of any possible effects on the health and safety of your employees – including the effects of noise emissions. Under PUWER you must provide information, instruction and training to your employees on the potential risks to their health and safety when using the equipment, and the precautions that may be necessary. You must also make sure that any equipment provided complies with any regulations which implement European Community directives, including the Supply Regulations.

242 You also have a duty under the Noise Regulations to reduce the risk to your employees from noise exposure. This means you need to take account of noise when selecting and providing tools and machinery for your employees.

243 You need information from the supplier or manufacturer to alert you to the noise produced by machines, to help you select suitable products and design the work processes for which they will be used, and to help you to plan your arrangements to protect your employees. You should consider the available information on noise emissions before you buy or hire tools or machinery. Information on noise emissions must be included in any technical documents describing the machinery, so you should be able to obtain this from the supplier.

244 Selecting quieter tools and machinery for your employees to use may take some time to bring benefits. Many employees are exposed to noise from multiple sources, so the initial work in selecting quieter machinery may not appear to be beneficial. Over time, as more quieter machinery is brought into use, the benefits of the initial work will be realised.

Duties of manufacturers and suppliers

245 The Supply Regulations place duties on manufacturers and suppliers of machinery. These are described in more detail in Appendix 4. For new machinery and any machinery first put into use after 1993 you can expect a manufacturer and/or supplier to:

■ design and construct the machinery so that risks resulting from noise emissions are reduced to the lowest level taking account of technical progress, in particular looking to reduce noise at source;

■ provide information to warn you where there are risks from noise which have not been eliminated ('residual risks');

■ provide information on the noise produced by the tool or machine at the operator's positions, both in terms of average noise levels and peak noise levels, and for the noisier machines provide information on the sound power emitted. Sound power is a measure of the total sound produced by the machine;

■ provide information on any specific requirement for training of the operator to ensure that low noise exposures are achieved and sustained, and any requirements for training those who will undertake maintenance of machines.

246 The Outdoor Equipment Regulations place duties on manufacturers and suppliers of certain classes of machinery commonly used outdoors to:

■ provide information on the sound power emitted by the machine;

■ for some of the machines covered by the Regulations, meet set targets for the sound power emitted.

Using manufacturers' noise data

247 Noise data supplied by manufacturers of tools and machines can indicate how much noise an operator of that machine is likely to be exposed to, and also how much noise the machine is likely to emit into the surrounding area. This can help to:

■ compare the noise from different brands and models of the same type of tool or machine;

■ identify (and avoid) tools and machines that have unusually high noise emissions;

■ consider the differences in noise of several tools/machines which are, in other respects, suitable for the particular task;

■ identify the range of likely noise levels when a tool/machine is used for different tasks or working on different materials;

■ make an estimate of noise exposure resulting from the use of the machine to assess risk and evaluate the need for controlling exposure.

248 The manufacturer's noise emission data should come with a description of the operating conditions under which the noise was tested. They should report the test procedures they have adopted, including the machine configuration and the operating and loading conditions during the test. Manufacturers are free to decide

the operating conditions, or they may choose to follow a standardised noise test code for the specific class or type of machine, where such a standard test exists.

249 Some general standard test codes are available to manufacturers to determine noise emission. These include:

- BS ISO 230-5:2000 *Test code for machine tools. Determination of the noise emission*;[12]

- ISO 7960:1995 *Airborne noise emitted by machine tools. Operating conditions for woodworking machines*;[13]

- BS EN ISO 9902-1:2001 *Textile machinery. Noise test code. Common requirements*;[14]

- BS EN 1265:1999 *Noise test code for foundry machines and equipment.*[15]

250 The declared noise emissions should include an indication of the 'uncertainty' in the declared value. The uncertainty is a factor designed to account for how repeatable the noise test is, and variations within the production process. Manufacturers should either declare the measured noise emission and the uncertainty as separate values, or should declare a single value that is the sum of the measured noise emission and the uncertainty. In the second case it is good practice for the manufacturer to state the uncertainty value separately. You should check whether any declared noise emission values include an uncertainty factor.

251 Values for uncertainty will typically be up to 3 dB, or higher for some types of machine. The difference between the noise emission values for two tools or machines should not be considered significant if it is smaller than one of the quoted uncertainty values. Note that it may not always be essential or desirable to choose the tool or machine with the lowest declared noise emission, but it must be safe and suitable for the particular task – you should aim to avoid tools or machines with above average noise.

252 Where a noise test does not produce emission values that adequately reflect real use of the tool or machine, the declared noise emission may not be sufficient to warn of the risk that the user must manage, and supplementary information is required. The manufacturers' instructions must inform you of residual risks, methods for safe use and, where necessary, training instructions.

253 Many European standards within which the noise tests are defined are currently under review and the revised versions may result in noise emission values that provide a more accurate guide to likely noise emissions during intended use. The standards may also in the future set out achievable levels of noise emissions for certain classes of tool or machine.

Limitations of manufacturers' noise data

254 At present, some of the noise test methods used by manufacturers do not represent the way tools or machines perform at work, and noise levels in the workplace may be higher than those obtained in this type of 'laboratory' test. This means that the manufacturer's declared noise emission values may not be representative of real use of the tool. For example:

- the operating conditions specified in the noise test may be a 'no-load' condition, ie no material is worked by the machine during the test;

- the noise test may specify a method which does not include all noise-generating mechanisms present during real use;

- the material being worked during the noise test may be different to that which the user intends to work;

- the manufacturer may have used a low-noise consumable or tooling during the noise test, which the end user may, for whatever reason, choose not to use.

255 Unless you are satisfied that the manufacturer's data on noise emissions reflect the real working conditions, and is representative of the work to which you intend to put the tool or machine, you should not use the emission data as a means of evaluating the noise exposure of your employees. However, you may still be able to use the data to compare machinery of the same type, and to identify the noisier or noisiest machinery, as long as the machinery has been tested under the same or comparable conditions.

256 Even when declared noise emissions do reflect noise emitted under real use, you should be cautious in using the data to predict noise exposures. This is because a worker's noise exposure often comes from a number of sources of noise, and may not be dominated by the noise from the tool or machine under consideration. You will need to assess the effect of the noise from the tool or machine in combination with other noise sources.

Residual risks

257 If there is a residual risk, after all practicable means of noise reduction have been incorporated in the design and construction of the machinery, the manufacturer must provide information so that you can put the machinery to use safely for all reasonably foreseeable applications. For example, the manufacturer of a hand-held grinding machine will need to consider the range of noise levels likely to be generated by his machine when used with various types of abrasive disc and the materials likely to be worked. He can then provide enough information to allow you to assess and manage the risk when operating the machine.

258 The declared noise emission value will often be enough to alert you to the need to control the noise risk, but where the test code does not produce realistic noise values, additional information is required to allow the equipment to be used safely (eg by specifying maintenance programmes, operating techniques, training requirements or likely in-use noise levels during the full range of intended uses of the machine).

Training requirements

259 Where tools or machines require specific training of the operator to ensure that low noise exposures are achieved and sustained, or training of others such as those who will undertake maintenance of machines, then you should expect the suppliers to alert you to this. For example, this might include:

- training in new operator skills for tools or machines with noise-reduction features;

- notification of applications of the tool or machine that produce unusually high noise emissions;

■ information about particular methods of using the tool or machine to be adopted or avoided that greatly affect the emitted noise; and

■ training in maintenance requirements to avoid unnecessary exposure.

Second-hand equipment

260 The Supply Regulations apply to all relevant machinery first supplied or put into service in the EEA from 1993. For machinery first supplied before 1993, section 6 of the Health and Safety at Work etc. Act 1974 requires importers and suppliers of machinery and equipment for use at work to ensure, so far as is reasonably practicable, that it is safe and without risks to health at all times when it is being set, used, cleaned or maintained by a person at work. They must also provide adequate information on the use for which the equipment is designed so that it can be used safely and without risk to health. Also, under regulation 10 of PUWER, machinery you provide for use at work by your employees must comply with the 'essential health and safety requirements' in the Supply Regulations.

261 Suppliers of second-hand machinery may be able to rely on information originally supplied with the machine if this is available and sufficient. However, they may need to provide new information if, for example, the original information is no longer available, if the machine has been significantly modified, so that the existing information is no longer valid, or if the original information did not meet the standard required.

Summary for employers: Selecting low-noise tools and machinery

262 When selecting equipment, besides ensuring that the tool or equipment is generally suitable for the job, you should:

■ ask about likely noise levels for your intended use(s) before making your final choice;

■ check that manufacturers' noise data is representative of likely noise levels for your intended use(s);

■ look for warnings in the instruction book to see if particular uses of the tool or machines are likely to cause unusually high noise;

■ be aware that even where manufacturers declare that their tool or machines produce less than 70 dB, levels may sometimes be much greater in your workplace.

PART 5: HEARING PROTECTION – SELECTION, USE, CARE AND MAINTENANCE

Overview

- What types of hearing protectors are available and what are they suitable for?

- How do I select protectors which give the right amount of protection, taking account of the work being done?

- How do I choose the right protection for the working environment, taking account of my employees' needs?

- How do I make sure protectors are compatible with other safety equipment?

- How should hearing protectors be used?

- What are the proper ways of fitting and wearing hearing protectors?

- How do I keep hearing protectors in good condition?

263 Providing personal hearing protection should be one of your first considerations on discovering a risk to the health of your employees due to noise. It should not be used as an alternative to controlling noise by technical and organisational means, but for tackling the immediate risk while other control measures are being developed. In the longer term, it should be used where there is a need to provide additional protection beyond what has been achieved through noise control. Where personal hearing protection is needed it is important that you select the right type of protection, and make sure that it is used and looked after.

264 Much of the information here is based on British Standard BS EN 458:2004 *Hearing protectors. Recommendations for selection, use, care and maintenance. Guidance document,*[16] which you may wish to consult for more detailed guidance.

265 Hearing protectors are available in many forms. They are all capable of providing a reduction in noise exposure and will be provided with information to allow you to decide whether they provide adequate noise reduction for your work situation. Whichever type of protector is used, it will provide its best protection only if it is in good condition, is the correct size and is worn properly.

266 All hearing protection should carry the CE mark. This is an indication that it meets a set of essential requirements, in accordance with the Personal Protective Equipment Regulations 2002.[17]

What to consider when selecting hearing protection

267 The following factors are likely to influence your selection of hearing protection:

- Types of protector, and suitability for the work being carried out.

- Noise reduction (attenuation) offered by the protector.

- Compatibility with other safety equipment.

- Pattern of the noise exposure.

- The need to communicate and hear warning sounds.

- Environmental factors such as heat, humidity, dust and dirt.

- Cost of maintenance or replacement.

- Comfort and user preference.

- Medical disorders suffered by the wearer.

268 Your choice of protectors may also be influenced by factors relating to intrinsic safety and electromagnetic compatibility, which are not covered in this book.

Figure 24 Types of earmuff

Figure 25 Earplugs with neck cord

Figure 26 Types of earplug

Figure 27 Earplugs with neckband

Figure 28 Custom-moulded earplugs

Figure 29 Helmet-mounted earmuffs

Table 5 Types of hearing protector, advantages, disadvantages, care and maintenance

Type	Description	Advantages	Notes on suitability and use	Care and maintenance
Earmuffs	Hard plastic cups which fit over and surround the ears, and are sealed to the head by cushion seals filled with a soft plastic foam or a viscous liquid. Tension to assist the seal is provided by a headband. The inner surfaces of the cups are covered with a sound-absorbing material, usually a soft plastic foam. Available in a range of sizes.	Easy to fit and use. Clearly visible therefore easily monitored.	Headband can prevent use of a hard hat. Headband can be worn behind the neck or under the chin if an under-hat support strap is provided. However, the protection offered may be reduced. May not be suited for use with safety glasses and other forms of personal protective equipment (check compatibility). May be uncomfortable in warm conditions. Long hair, beards and jewellery may interfere with seals and reduce protection.	Check seals for cleanliness, hardening, tearing and misshape. Check cup condition for cracks, holes, damage and unofficial modifications. Avoid over bending or twisting headband, which may degrade performance. Check tension of headband (compare with a new earmuff). Store in a clean environment. Follow manufacturer's instructions.
Helmet-mounted earmuffs	Individual cups attached to safety head-gear such as a visor or a hard hat, usually by adjustable arms. Noise protection information should be obtained for the specific combination of earmuff and helmet.	Can overcome the difficulties with compatibility with hard hats.	May not be suited for use with safety glasses and other forms of protective equipment (check compatibility). May be uncomfortable in warm conditions. Long hair, beards and jewellery may interfere with seals and reduce protection.	As for earmuffs. Ensure the seals do not sit on the side of the helmet for long periods as this can damage them and affect their performance.
Earplugs	Earplugs fit into the ear or cover the ear canal to form a seal. They sometimes have a cord or a neckband to prevent loss. Some plugs are reusable and others are designed to be disposed of after one use – check manufacturer's instructions. Available in different forms (pre-shaped, user-formable, semi-insert).	Suited for use with safety glasses and other forms of personal protective equipment.	Can be hard to fit – will only be effective when fitted properly so correct fitting is essential. See manufacturer's instructions and provide training. Difficult to check correct fit by observation. Can work loose over time, so allow for refitting in a quiet environment every hour. May not be suitable where the hearing protection is likely to be removed often, particularly in dusty or dirty environments. May not be suitable for certain individuals due to medical conditions.	Clean reusable plugs regularly and ensure they are not damaged or degraded – follow manufacturer's instructions. Disposable earplugs should only be used once. Hands should be clean when fitting earplugs. Reusable plugs should be issued to an individual and not shared. Ensure adequate supplies of disposable earplugs. Follow manufacturer's instructions.

Type	Description	Advantages	Notes on suitability and use	Care and maintenance
Custom-moulded plugs	Earplugs made from a material such as silicone rubber, individually moulded to fit a person's ears.	For some wearers they may be easier to fit than other types of earplug and therefore more likely to get good protection. Comfortable.	Poor performance if manufacturing and initial fitting are not done properly. Ensure fit tests are done before plugs are put into use.	As for earplugs.

Figure 30 Earmuffs worn with headband behind head (note use of support strap)

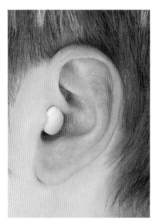

Figure 31 Correct fitting of earplugs

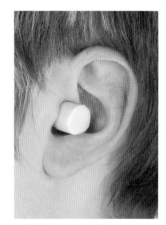

Figure 32 Incorrect fitting of earplugs

Figure 33 Problems of fitting earmuffs with long hair

Figure 34 Problems of fitting earmuffs with jewellery

Figure 35 Problems of fitting earmuffs with safety glasses

Special types of protector

Level-dependent protectors

269 Level-dependent (or amplitude-sensitive) hearing protectors are designed to protect against hazardous noise while permitting good communication when it is quieter. They are most suited to situations where the noise exposure is intermittent and there is a need to communicate during quieter intervals.

270 Sound-restoration, level-dependent hearing protectors are available, incorporating an electronic sound-reproduction system. At low sound pressure levels, the sound detected by an external microphone is relayed to the inside of the hearing protector. As the external sound level increases, the electronics gradually reduce the transmission of sound.

271 Level-dependent devices based on non-electronic methods are also available. These use the acoustic properties of carefully designed air ducts to give different protection at different noise levels. These types of protector are designed to be effective against very high single-impulse noises, such as firearms, rather than the continuous noise or repetitive impulses found in most industrial situations.

Flat or tailored frequency response protectors

272 Whereas most hearing protectors provide greater reduction of noise at high frequencies than they do at low frequencies, this type of protector, by its design, gives a similar reduction across a wide frequency range (ie a flat frequency response). This can assist effective communication, and can be useful in circumstances where it is important to be able to hear the high-frequency sound at the correct level relative to the low-frequency sounds, eg musicians during rehearsal and practising.

273 Similarly, hearing protection is available which is designed to reduce low- and high-frequency noise in particular, potentially allowing a greater degree of spoken communication.

Active noise-reduction protectors

274 Active noise-reduction (ANR) hearing protectors incorporate an electronic sound cancelling system to achieve additional noise reduction. ANR can be effective at low frequencies (50-500 Hz) where ordinary protectors can be less effective. ANR protectors are usually based on an earmuff type protector.

Protectors with communication facilities

275 These devices make use of wired or aerial systems to relay signals, alarms, messages or entertainment programmes to the wearer. These protectors should be designed so that the level of the relayed signal is not too loud. Where the devices are used to receive spoken messages the microphone should, where possible, be switched off when not in use, to avoid the reproduction at the ear of spurious background noise.

276 When considering these devices, check to ensure it is possible to hear necessary warning sounds (eg speech and safety alarms)* above the sounds reproduced at the ear. Safety alarms should not normally be relayed through the communication system because of the risk of system failure.

* Standard methods are available for selecting suitable auditory warning signals - BS ISO 7731:2003 *Ergonomics. Danger signals for public and work areas. Auditory danger signals.*

Care and maintenance

277 Hearing protection must be monitored for wear and damage and replaced when necessary. If hearing protectors are to be effective, and provide the expected protection, they must be in good condition. You are responsible for ensuring that hearing protection is well maintained, while employees are responsible for reporting any defects (see regulation 8). With experience, simple checks can be made by visual inspection and feel. It is good practice to keep a set of new protectors on display, to provide a basis for comparison.

Figure 36 Damaged earmuff seal

Figure 37 Damaged helmet-mounted earmuff seal due to prolonged storage in 'up' position

Figure 38 Earmuff headband showing reduced tension (in comparison to tension when new)

Likely noise reduction

Reduction of noise exposure

278 The Noise Regulations require that hearing protection is selected to eliminate the risk to hearing, or reduce the risk to the lowest level reasonably practicable and that the selection process takes account of consultation with employees or their representatives. You should aim to provide protection that at least reduces the A-weighted sound pressure level at the wearer's ear to below 85 dB.

279 For impulsive noise, you should aim to provide protection sufficient to reduce the C-weighted peak sound pressure level at the ear to below the upper peak exposure action value of 137 dB. The noise reduction can be estimated using the method given in Appendix 2.

280 For workers with variable exposures, you should make sure that your employees have protectors adequate for the worst situation likely to be encountered and that they know when and where to use them.

281 Three methods for predicting the protection provided by hearing protection devices are described in Appendix 3. You should use one of these methods to establish the predicted protection, and before going on to account for 'real world' factors as described in paragraphs 282-286.

'Real world' protection

282 It is very likely that under conditions of real use, hearing protectors will give lower protection than predicted by manufacturers' data which is obtained from standardised tests. The difference between manufacturers' data and 'real-world' protection is due to factors such as poor fitting and wearing of spectacles or other personal protective equipment. You should account for this 'real world' protection by 'derating' the protector by 4 dB. This means, having followed one of the methods in Appendix 3, assume that the level at the ear when hearing protection is worn will be 4 dB higher than would be predicted by the method. In this way you will get a better indication of the protection that users are most likely to get, and can select appropriate hearing protection accordingly.

283 The derating does not apply to the assessment of hearing protector performance against peak noise, which is described in Appendix 2.

284 The 4 dB derating described in paragraph 282 is regarded as an appropriate factor to bridge the gap between manufacturers' data and real-world factors, without introducing further complexity to the prediction of hearing protector performance. You will still be able to select an appropriate hearing protection device for the character of the noise, and hearing protectors that show better repeatability in standardised tests will still be distinguishable.

285 The use of a derating factor will not necessarily mean that you will need to select a protector with a higher rating than one you currently use. By following all the guidance within this Part you should be able to ensure that even if wearers get better protection than that predicted by these methods, the protected level at the ear will be within the recommended range.

286 You may wish to demonstrate by means other than relying on manufacturers' data and the methods outlined in this book that the hearing protection you supply meets the requirement to reduce the noise level at the ear to the appropriate level, eg by measurement of noise levels underneath hearing protection devices in conditions of real use.

Over-protection

287 Protectors that reduce the level at the ear to below 70 dB should be avoided, since this over-protection may cause difficulties with communication and hearing warning signals. Users may become isolated from their environment, leading to safety risks, and generally may have a tendency to remove the hearing protection and therefore risk damage to their hearing.

288 Table 6 gives an indication of the protector factor that is likely to be suitable for different levels of noise. It based on the single number rating (SNR) value provided with a hearing protection device (see Appendix 3). The information is intended as a guide rather than a substitute for using one of the methods in Appendix 3, and in particular will not be appropriate if there are significant low-frequency components to the noise in question. Examples of noise environments which may contain significant low-frequency components, and for which this table is not suitable, include press shops, generators and generator test bays, plant rooms, boiler houses, concrete shaker tables, moulding presses and punch presses.

Table 6 Indication of protector factors

A-weighted noise level (dB)	*Select a protector with an SNR of . . .*
85-90	20 or less
90-95	20-30
95-100	25-35
100-105	30 or more

Dual protection

289 People working in extreme noise conditions may require more protection than that provided by earmuffs or earplugs alone. This problem is likely where the daily noise exposure is above 110 dB or the peak sound pressure level exceeds 150 dB, especially if there is substantial noise at frequencies less than about 500 Hz.

290 Improved protection can be obtained by wearing a combination of earmuffs and earplugs. The amount of protection will depend on the particular earmuff and plug combination. In general, the most useful combination is a high performance plug with a moderate-performance earmuff (a high-performance earmuff adds a little extra protection but is likely to be less comfortable).

291 If dual protection is used, test data should be obtained for the particular combination of plug and earmuff (and helmet, if used). In practice, the increase in attenuation you can expect from dual protection will be no more than 6 dB over that of the better of the individual protectors.

Other factors influencing selection

Wearer comfort and preference

292 Individuals differ in what they find comfortable. Some people prefer earplugs in hot environments, but others find any earplug extremely uncomfortable and prefer earmuffs. Wherever possible you should make more than one type of protector available (making sure that each is suitable for the noise and the jobs to be done) to allow the user a personal choice.

293 All protectors are likely to be somewhat uncomfortable, especially in hot, humid conditions. Therefore, choose hearing protection that is sufficient to control the risk, does not over protect and is reasonably comfortable to wear.

294 One of the factors affecting the comfort of earmuffs is the pressure of the seals on the head. This can be kept low by using resilient seals which only need a low headband force. A high contact area between seal and head also helps reduce the contact pressure, but in hot conditions can cause the skin to sweat* – in these conditions earplugs may be preferred. Other important factors affecting comfort include the weight of the earmuffs (the lighter the better) and the size of the cup (the cups must be large enough to fit right over the user's ear).

Pattern of noise exposure

295 Where patterns of noise exposure are likely to be repeated and short-term, earmuffs and semi-aural/semi-insert earplugs may be preferred because they are quick and easy to fit and remove, and therefore more likely to be fitted when exposure occurs.

Warning signals and speech communication

296 Where it is important that certain sounds within the workplace are heard clearly and distinctly, such as with speech communication, warning signals and other informative sounds, hearing protectors with a flat frequency response may be preferred.

297 There is some evidence that the wearing of hearing protectors can increase speech intelligibility against a background of noise. You should not discount the use of hearing protection simply because there is a need for speech communication.

Location of a sound source

298 Hearing protectors can impair the wearer's ability to identify the location of a sound source. If this ability is necessary or important for your work activity then earplugs may be preferred since with earplugs the impairment is less than with earmuffs.

Compatibility with other safety equipment

299 When selecting hearing protectors you should take account of any other personal protective equipment that the user must wear which may impair the performance of the hearing protectors you have selected. Particular examples are the use of safety glasses or goggles, the frame or band of which may interfere with earmuff seals and reduce the protection offered. Where there is a need to wear hearing protectors in combination with helmets or face-shields, space may be limited and earplugs or low-mass earmuffs may be preferred.

Medical disorders

300 Part of the selection process for hearing protectors includes finding out whether the user has any medical disorder that could influence the selection. Medical disorders can mean any type of earache, irritation of the ear canal, discharge, hearing loss, or any type of ear disease or skin disorder. Where you have employees who have any such medical disorders you should seek medical advice as to the suitability of hearing protection.

---

* Liners which fit between the seal and the head can absorb sweat, but may reduce protection by a small amount (typically 2 to 4 dB).

Advice on issuing hearing protection

301 When issuing hearing protectors you will need to consider regulations 7 (hearing protection), 8 (Maintenance and use of equipment) and 10 (Information, instruction and training) of the Noise Regulations (see Part 1) and take account of the points below.

Information

302 You should provide your employees with information on:

- why you are issuing hearing protectors;

- where they must be used;

- how they can obtain replacements or new protectors; and

- how they should wear them properly and look after them.

Ready availability

303 You should ensure that employees can readily obtain hearing protectors and replacements when they need them. This might include personal issue to individual employees. Alternatively, you might wish to install dispensers from which employees can take disposable hearing protectors as they need them. Locate the dispensers at or near to the entrances to areas where hearing protectors are required, in a place where your employees can conveniently use them. Make sure you keep the dispensers topped up.

Personal issue and visitors

304 People should not pass earplugs to one another. Preferably, a set of earmuffs should be used by one individual only. Where earmuffs are kept for the use of visitors, they should be hygienically cleaned for each new wearer. Alternatively, disposable covers may be used.

Training and effective use

305 Hearing protection will only provide good protection when used properly and fitted correctly. Users must be instructed in its correct fitting and use, including:

- how to avoid the potential interference of long hair, spectacles and earrings on the effectiveness of their hearing protection;

- how to wear their hearing protection in combination with other personal protection;

- the importance of wearing their hearing protection at all times in a noisy environment (removing it for only a few minutes in a shift will lower the protection to the wearer considerably);

- how to store their hearing protection correctly;

- how to care for and to check their hearing protection at frequent intervals;

- where to report damage to their hearing protection.

306 This training may be provided by a suitably trained supervisor.

307 Some people tend to remove hearing protectors when speaking to others in noisy environments. You should advise them not to do this and explain to them that once they are used to the situation they will be able to communicate more easily with protectors than without them. Advise them to speak 'to the protector', ie to speak with the mouth close to the ear of the person to whom they are talking.

308 Some people tend to speak quietly when they are wearing hearing protectors in noisy areas because they can hear their own voice more clearly. This can cause communication problems, so you should advise users to speak up when wearing protectors.

Maximising performance of protectors through full use

309 There are many reasons why hearing protectors give less noise reduction than would be predicted. One of the most common reasons is that protectors are not used all of the time in noisy areas. If the protectors are removed in noisy areas, even for short periods, the amount of protection provided will be severely limited. Employers should ensure, through training and proper supervision, that employees wear their hearing protectors at all times when they are required. Employees have a duty to make full use of hearing protectors which have been provided to them.

310 Figure 39 shows the how the effectiveness of a hearing protector is reduced if it is not worn all the time it should be. It shows the effective protection offered by three different hearing protectors against the percentage of time worn. When the protectors are worn for 100% of the time that the user is exposed to the noise, they give the expected protection. As wear time is decreased the effective protection offered decreases. A significant reduction in protection is found even if the wear time is 90%. If the protectors are worn for 50% of the time they should be, the protection offered is only about 3 dB.

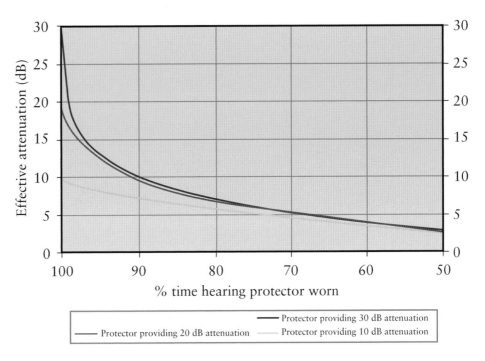

Figure 39 Effectiveness of hearing protection in relation to time worn

PART 6: HEALTH SURVEILLANCE FOR HEARING DAMAGE

Overview

■ What is health surveillance and why is it necessary?

■ What do I have to do?

■ What are the principles which apply in setting up programmes of hearing checks?

What is health surveillance?

311 Health surveillance is about putting in place systematic, regular and appropriate procedures to detect early signs of work-related ill health, and acting upon the results. The aims are primarily to safeguard the health of workers (including identifying and protecting individuals at increased risk), but also to check the long-term effectiveness of measures to control risks to health.

312 Health surveillance for noise-induced hearing loss (NIHL) usually means regular hearing checks (audiometric testing) to measure the sensitivity of hearing over a range of sound frequencies. It should include informing employees about the state of their hearing and the keeping of records.

When is health surveillance required?

313 You should provide health surveillance to workers regularly exposed above the upper exposure action values. Where exposure is between the lower and upper exposure action values, or where employees are only occasionally exposed above the upper exposure action values, health surveillance will only be required if information comes to light that an individual may be particularly sensitive to NIHL. This may be from past medical history, audiometric test results from previous jobs or other independent assessments. A few individuals may also indicate a family history of becoming deaf early on in life. If this information is brought to the attention of the employer then those individuals should be provided with health surveillance.

314 When deciding which workers will require health surveillance you should consider patterns of work. You would not be expected to provide health surveillance, for example, to workers whose noise exposure exceeds the upper exposure action values by a small margin, and only on one or two days a week. However if exposure is regularly above the upper exposure action values then health surveillance would be appropriate.

What do I actually have to do?

315 Health surveillance for NIHL will involve a system of audiometric testing. Full details of how to conduct the testing are in Appendix 5. You will need to appoint a person or people to be in charge of the testing programme. These people should be fully conversant with the technical and ethical aspects of the conduct of audiometry, and in particular be responsible for:

■ the quality of the service provided;

- ensuring that appropriate standards are maintained during testing;

- record-keeping;

- referring individuals for further advice.

316 A suitable person might be an occupational physician or nurse with training in audiometry, an audiological scientist or a trained audiometrician with access to a qualified occupational health medical professional for advice and onward referral. The provision of health surveillance may be contracted to an external occupational health service provider. Ultimately you as the employer have the responsibility for the proper conduct of health surveillance. The detailed arrangements for further advice should be agreed between you and the person in charge of the testing programme.

317 The person performing the tests should have appropriate training so that testing is carried out in a repeatable and accurate manner. A training syllabus for industrial audiometricians has been prepared by the British Society of Audiology (www.thebsa.org.uk) which has approved a number of courses. The basic requirements for any person responsible for the testing programme are that they should:

- have a good understanding of the aims and objectives and technique of audiometry and how it relates to hearing;

- be able to carry out otoscopic examination of the ear to detect any major abnormality or the presence of discharge or wax which might affect the results;

- be competent to ensure an appropriate test environment, to operate and maintain the testing equipment, and to carry out the test procedure;

- understand and comply with the need for confidentiality of personal health information;

- know how to assess and present results according to a defined system, and when and how to seek medical assistance.

318 In addition the tester should be familiar with the hearing protection used by the workers to discuss proper fitting, cleaning and maintenance.

319 The best programme would begin with a baseline audiogram, giving details of the condition of an individual's hearing before exposure to noise. A baseline audiogram may be a pre-employment test or the first audiogram to be conducted on taking up employment or moving to a noisy job. The benefits of a pre-employment test are that it precludes any exposure to noise within your employment. A programme can, however, be introduced at any time for employees already exposed to noise. This would be followed by a regular series of tests, usually annually for the first two years and then at three-yearly intervals, although testing may be more frequent where an abnormality in hearing is detected or where the risk of hearing damage is high.

320 The person conducting the testing should explain the results of each test to the employee including the condition of the individual's hearing, the significance of hearing damage, the importance of following your noise-control and hearing-protection programmes and what happens next if any abnormality in hearing is detected. If the test is not conducted by a doctor then the employee may need to be referred on to their GP if damage is found.

321 Analysis of the results of health surveillance for groups of workers can provide an insight into how well your hearing conservation programme is working. You should use these results to provide feedback to your risk assessment and your noise reduction, education and compliance programme. You should also make this information available to employee or safety representatives.

322 If the work is contracted out to an external provider you should ensure they are competent and agree the terms of the contract including procedures for feedback of grouped results and record-keeping. In these circumstances more effort may be required to ensure you receive appropriate feedback of results of testing.

323 So that your employees understand what health surveillance means to them, you should discuss your programme with them and their representatives at an early stage. You should consult employee or safety representatives in good time about the development of the programme. The following issues should be covered in your consultation and discussions:

■   the aims and objectives of the programme;

■   how the programme fits with the more general aim of preventing hearing damage at work;

■   the procedures to ensure confidentiality of the results;

■   the methods to be followed, including those for medical referral;

■   how the programme will be monitored and evaluated.

What type of records should I keep?

324 You should keep an up-to-date health record for each individual as long as they are under health surveillance. These records should include:

■   identification details of the employee;

■   the employee's history of exposure to noise;

■   the outcome of previous health surveillance in terms of fitness for work, and any restrictions required.

325 Health records should not contain personal medical information, which must be kept in confidence in the medical record held by an occupational health professional in charge of the audiometric testing programme. If the person in charge of the programme is not medically qualified then the holding of medical information should be agreed between you, your employees and their representatives at the set up of the programme. This means that the person in charge of the testing programme should hold the results of testing independently from you and the personnel records and abide by the principles of confidentiality normally expected of health professionals. This person should forward the medical information to the GP or consultant where problems with hearing are identified.

326 You will need to see anonymised grouped data on the hearing of the workforce. This should be done in a way that does not reveal details of any particular individual's hearing threshold and does not compromise confidentiality. Consent will not be required for this type of information to be provided to you. It may be difficult to provide grouped anonymised data for small groups of workers.

However, ways should be devised to inform management of issues that need to be dealt with to prevent hearing loss without compromising confidentiality.

327 The health record should be retained for at least as long as the employee remains in your employment. You may wish to retain it for longer as enquiries regarding the state of an individual's hearing may arise many years after they have left your employment or exposure to noise has ceased. It is good practice to offer individual employees a copy of their health record when they leave your employment. Inspectors from the enforcing authorities are entitled to ask to see your health records as part of their checks that you are complying with these Regulations. If your business should cease trading you may wish to pass the record to the individual concerned.

Appendix 1

Measuring noise exposure in the workplace

Overview

■ What should be measured?

■ What instruments can be used?

■ What are the general principles, techniques and strategies that should be adopted when making measurements?

What should be measured?

1 When making measurements to estimate the noise exposure of a person at work, you need to ascertain the equivalent continuous A-weighted sound pressure level (L_{Aeq}) that represents the noise the person is exposed to during the working day. You also need to ascertain the maximum C-weighted peak sound pressure level or levels to which the person is exposed.

2 The L_{Aeq} is combined with the duration of exposure during a working day to ascertain the daily personal noise exposure, $L_{EP,d}$, using the formula defined in Schedule 1 Part 1 paragraph 1 to the Noise Regulations.

3 In practice it is common to break the working day into a number of discrete jobs or tasks, and to make sample measurements to determine a representative L_{Aeq} for the job or task. The L_{Aeq} for each job or task is then combined with its duration during the working day to ascertain the $L_{EP,d}$, using the formula defined in Schedule 1 Part 1 paragraph 2 to the Noise Regulations.

4 Electronic spreadsheets are available on the HSE website (www.hse.gov.uk/noise) which allow the calculation to be performed. A simple 'ready-reckoner' method for determining daily personal noise exposure using these measurement parameters is described in Part 2.

5 You may also wish to make other types of measurements, such as the equivalent continuous C-weighted sound pressure level (L_{Ceq}) or the L_{eq} in octave frequency bands, for a job or task to perform calculations to predict the performance of personal hearing protection.

Instruments

What do I use to measure noise?

6 The basic instrument for measuring noise is a sound level meter. A dosemeter (personal sound exposure meter) worn by the employee can also be used. Dual-purpose instruments are also available which can operate as both a sound level meter and a dosemeter. A calibrator to check the meter and a windshield to protect the microphone against air movement and dirt are essential accessories.

7 Other, more sophisticated, equipment such as data recorders, frequency analysers, and sound intensity analysers can be used for a more detailed assessment. This equipment is not covered by this book.

Sound level meters

8   The sound level meter should be an integrating sound level meter capable of measuring the basic parameters described in paragraph 1 and, optionally, the parameters described in paragraph 5.

9   Your sound level meter should meet at least Class 2 of BS EN ISO 61672-1:2003[18] (the current instrumentation standard for sound level meters), or at least Type 2 of BS EN 60804:2001[19] (the former standard).

Personal sound exposure meters (Dosemeters)

10   Where a person is highly mobile or working in places where access for the measurement is difficult, a dosemeter is an alternative means of measuring a person's noise exposure.

11   Dosemeters indicate the total noise dose received over the measurement period. Modern dosemeters commonly indicate the L_{Aeq} over the measurement period. Some meters indicate the dose in units of Pascal squared hours $(Pa^2.h)^*$ or as a percentage of a given $L_{EP,d}$ (usually 85 or 90 dB). Meters are required to provide a means of converting the reading to $Pa^2.h$ if this is not directly indicated on the meter.

12   In the case where the meter indicates the dose as a percentage of an $L_{EP,d}$ there may be an assumption that the measurement period corresponds to the whole working day, or there may be the ability to key in a value for the length of day so that the instrument can make the calculation. You should make sure you understand how the $L_{EP,d}$ shown by the meter is calculated.

13   Many dosemeters have additional features. Those which record how the sound pressure level varies with time throughout the measurement (a logging dosemeter) can be useful to show when and where high noise exposures occur.

14   All dosemeter measurements should be made with a 3 dB exchange rate (sometimes called the doubling rate).

15   People wearing dosemeters should be instructed not to interfere with the instrument or microphone during the course of the measurements. They should also be instructed not to speak more than is necessary during the course of the measurement, since a person's own voice should not be included in an assessment of their daily personal noise exposure.

16   Your dosemeter should meet the requirements of BS EN 61252:1997.[20] Dosemeters have no type or class number.

Calibrators

17   A sound calibrator should be used to check the meter each day before and after making any measurements. Calibrators give a tone at a specified sound pressure level and frequency for a specified microphone type using an appropriate adaptor. Make sure you have the right calibrator with the right adaptor for your microphone.

18   Some meters have an internal electronic calibration. The internal calibration only checks the instrument's electronics and does not provide a check of the microphone. However, it can be a useful cross-check of the meter and calibrator.

* A measure of the total sound energy received during a measurement period.

19   Your calibrator should meet at least Class 2 of BS EN 60942:2003.[21]

Peak sound pressure level

20   Peak sound pressure should be measured with a C-weighting applied. You should ensure when measuring peak sound pressure that the correct frequency-weighting is applied. Some sound level meters include an 'I' (impulse) response. This should not be used for any measurements relating to the requirements of these Regulations.

21   More information on peak sound pressure and its measurement is in Appendix 2.

Periodic testing of instruments

22   Both your meter and calibrator need to have been tested in the previous two years to ensure they still meet the required standards. If your equipment is more than two years old, check you have a test certificate confirming the performance of your meter and calibrator before you start your assessment. More details on periodic testing are given in paragraphs 61-62 of this appendix.

Where should I measure and how should the measurements be made?

Where?

23   When measuring to estimate a person's noise exposure, make measurements at every location they work in or pass through during the working day, and note the time spent at each location. It is generally not necessary to record exposures to sound pressure levels below 75 dB, since such exposures are unlikely to be significant in relation to the daily noise exposure action levels.

24   Measurements should be made at the position occupied by the person's head, preferably with the person not present. Operators may need to be present while the measurements are made, eg to control a machine or process. Measurements should be made with the microphone positioned close enough to the operator's head to obtain a reliable measure of the noise to which they are exposed, but preferably not so close that reflections cause errors. The results are unlikely to be significantly affected by reflected sound if the microphone is kept at least 15 cm away from an operator. The microphone should be placed on the side of the head where the noise levels are higher.

Figure 40 Making measurements with a hand-held sound level meter

Figure 41 Recommended position for a dosemeter microphone

25   To avoid making large numbers of measurements, eg where the sound pressure level is changing, or if the person is moving within a noisy area, you may wish to assume the worst case and measure at the noisiest location, or during the loudest periods. Alternatively, carrying out a spatial-average measurement by following the movement of the worker may provide a representative measure of the noise exposure.

26   If you are using a dosemeter to measure a person's noise exposure, position the microphone on the shoulder (ideally on the shoulder joint) and prevent it touching the neck, rubbing on or being covered by clothing or protective equipment. If the dosemeter body is connected to the microphone by a flexible cable, place the meter securely in a pocket or on a belt where it can be safe from damage during the measurement.

How long to measure?

27   The noise level to which an individual employee is exposed will normally change throughout the day because, for example, different jobs might be done and different machines or materials might be used at different times. You must take enough noise measurements to account for all these changes, recording the sound level and the person's exposure duration at each noise level.

28   With a sound level meter, you need to measure at each position or during each job or task, long enough to obtain a representative measurement of the level the person is exposed to. You may need to measure the L_{Aeq} for the entire period but a shorter measurement can be sufficient. In general:

■   if the noise is steady, a short sample L_{Aeq} measurement may be enough;

■   if the noise is changing, wait for the L_{Aeq} reading to settle to within 1 dB;

■   if the noise is from a cyclic operation measure the L_{Aeq} over a whole number of cycles.

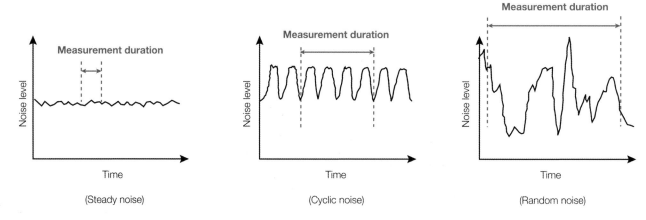

Figure 42 Representative measurement durations for different types of noise

29   The time required depends on the nature of the work and the characteristics of the noise. You should ensure by observation of the work that your measurement covers all significant noise during the job or task. In particular it is important to make sure that any short-duration, high-level noise exposures are included in the measurement, since these can have a significant impact on the true L_{Aeq}.

30   Noise dosemeters are designed to operate for long periods. They are ideal for measurements over an entire shift, or for a period of several hours during a shift. If you measure over part of a shift, make sure the period of your measurement covers all significant noise exposure, so as to be typical of the working day, so that you can reliably predict the full daily exposure. Avoid very short measurements which can be inaccurate due to the limited resolution of the dosemeter's display. Also make sure that the dose reading relates to actual true noise exposure, not false input from unrepresentative noise sources when the meter is not supervised, eg artificial bangs, whistling, blowing and tampering with the microphone.

Sample measurements for a group

31   If several workers work in the same area, you may be able to estimate the exposure for them all from measurements in selected locations. When making the measurements, choose the locations and times spent in each place so that you determine the highest exposure that someone is likely to receive.

Mobile workers and highly variable daily exposures

32   For some jobs (such as maintenance) the work and the noise exposure will vary from day to day so there is no typical daily exposure. For people in these jobs, measurements need to be made of the range of activities undertaken, possibly over several days. From these measurements the likely daily exposure for a nominal day or days should be estimated.

Measurements close to the ear

33   Measurements of noise very close to the ear, such as sound from a communication headset, or under a motorcycle helmet, require specialist equipment and procedures. Further details on measuring noise very close to the ear are given in paragraphs 49-56 of this appendix.

Sources of error and other factors influencing the measurement result

34   Sources of error should as far as possible be avoided. To reduce errors, it is important to distinguish between sources of error and natural variables. The major factors influencing the result are listed in Table 7.

Table 7 Factors contributing to the measurement result

Factor	Treatment
Impacts on microphone/cable	Error
Wind-induced noise	Error
Reflection from body to microphone	Error
Noise from PA systems, radio etc	Include in measurement
Speech (subject's own)	Aim to exclude from measurement
Speech (other people)	Include in measurement
Variations in local sound level	Natural variable
Noise from hand-held tools	Natural variable
Close-to-ear noise level	Natural variable
Duration of each operation	Natural variable

35   The relevant variables should be revealed during an analysis of the work under consideration and during measurements. If significant contribution from sources of error is detected, the measurements should be rejected or corrected.

36   The measured noise exposure and the uncertainty in the result depend on the measurement method used. A dosemeter tends to increase the potential false contributions to measurements and thereby the measured sound pressure level.

37   However, using a hand-held sound-level meter may lead to an underestimation of the worker's noise exposure. This is particularly connected to the difficulty in assessing the contribution from close-to-ear sound levels and noise from hand-held tools.

Using the results from dosemeter readings

38   As described in paragraphs 11-12, noise dosemeters may give results in terms of $Pa^2.h$, or as a percentage dose where 100% can correspond to an $L_{EP,d}$ of 85 or 90 dB. Paragraphs 39-42 show how to use these measurement parameters when estimating daily personal noise exposure.

39   Some dosemeters will give a result simply as an estimate of $L_{EP,d}$. This result will assume that the measurement period corresponds to the full working day; if this is not the case for your measurements you will need to obtain the result from the dosemeter in terms of $Pa^2.h$, percentage dose or L_{Aeq} to determine your $L_{EP,d}$.

40   **If the dose is given in $Pa^2.h$**, multiply the dose by 100 to obtain the 'noise exposure points' (see Part 2) for that dose.

41   **If the dose is given as a percentage**, then:

■   **if 100% corresponds to an $L_{EP,d}$ of 85 dB**, the 'noise exposure points' is the same as the percentage dose value;

■   **if 100% corresponds to an $L_{EP,d}$ of 90 dB**, multiply the percentage dose by 3.2 to obtain the 'noise exposure points' for that dose.

42   If the measurement period covers only part of the working day or of the exposure to noise, but is representative of the whole day or exposure, you can apply a multiplying factor to the exposure points calculated above to obtain the points for the exposure. The multiplying factor is:

$$\frac{\text{Duration of exposure}}{\text{Duration of measurement}}$$

Using results of dosemeter measurements – Worked example

An employee wears a dosemeter for $2\frac{1}{2}$ hours between breaks and the recorded dose is 1.3 $Pa^2.h$. The noise during the measurement period was typical for that work, but the employee is normally exposed to this noise for 6 hours per day.

Step 1   Multiply $Pa^2.h$ value by 100 to obtain noise exposure points for that dose:

1.3 x 100 = **130 points**.

Step 2	To obtain noise exposure points for the normal duration of exposure, multiply by exposure duration/measurement duration (6/2$^1/_2$):
	130 x 6/2$^1/_2$ = **312 points.**
Result	This task contributes 312 points to daily personal noise exposure.

Accounting for the contribution of peak noises to daily exposure

43   Where events such as impacts or impulses occur during the normal working day as part of the typical noise emission from a machine or process, they will contribute to a measurement of L_{Aeq}, as long as they have not been specifically excluded and the instrumentation used has sufficient dynamic range. There may be circumstances when it is necessary or desirable to assess, separately, the contribution of these events to a person's daily exposure. This can be achieved in two ways: measuring the sound exposure level (L_{AE}) for a single or known number of events, or measuring the L_{Aeq} over a known period which contains a single or known number of events. These methods are described in paragraphs 45-48.

44   It is important to ensure, if assessing the noise in this way, that you do not inadvertently account for the contribution from these events twice. You should make sure that the general assessment of noise exposure excludes the contribution from the events.

Assessing the contribution of events from sound exposure level (L_{AE}) measurements

45   The contribution to daily personal noise exposure from events such as impacts and impulses may be determined from a measurement of L_{AE} containing a known number of events, as long as the likely number of events per day is known.

46   The following equation may be used to determine the number of exposure points (EP) resulting from the events.

$$EP = \frac{n}{m} .10^{\left[\frac{L_{AE} - 109.6}{10}\right]}$$

where:

n is the number of events during the day; and

m is the number of events during the measurement.

Worked example of L_{AE} calculation

Proof firings are carried out of four shotgun cartridges. The sound exposure level L_{AE} is 122 dB. The operator would fire 20 cartridges per day.

Step 1	Substitute the following values into the equation:

$n = 20$

$m = 4$

$L_{AE} = 122$

Step 2	Calculate the exposure points

$$EP = \frac{20}{4}.10^{\left[\frac{122 - 109.6}{10}\right]} = 5.10^{1.24} = 87$$

Result	The 20 firings contribute 87 points to the daily personal noise exposure of the operator. Add these noise exposure points to those from other noise exposures during day, and determine total exposure as described in Part 2.

Assessing contribution of events from L_{Aeq} measurement

47   The contribution to daily personal noise exposure from events such as impacts and impulses may be determined from a measurement of L_{Aeq} over a known period containing a known number of events, provided the likely number of events per day is known.

48   The following equation may be used to determine the number of exposure points (EP) resulting from the events.

$$EP = \frac{n}{m}.T_e.10^{\left[\frac{L_{Aeq} - 109.6}{10}\right]}$$

where:

T_e is the duration of the measurement, in seconds;

n is the number of events during the day; and

m is the number of events during the measurement.

Worked example of L_{Aeq} calculation

A gamekeeper was exposed to noise when firing his rifle. Measurements were made at the side of his head with a sound level meter. The L_{Aeq} measured over a 30 second period when 3 shots were fired was 111 dB. During a day he would normally fire up to 25 shots.

Step 1   Substitute the following values into the equation:

$n = 25$

$m = 3$

$L_{Aeq} = 111$

$T_e = 30$

Step 2   Calculate exposure points

$$EP = \frac{25}{3}.30.10^{\left[\frac{111 - 109.6}{10}\right]} = \frac{25}{3}.30.10^{0.14} = 345$$

Result	The 25 firings contribute 345 points to the daily personal noise exposure of the gamekeeper. Add these noise exposure points to those from other noise exposures during day, and determine total exposure as described in Part 2.

Measurements of noise very close to the ear

49   When a person is receiving significant noise exposure from sources close to the ear such as communication headsets or earpieces, or they are wearing helmets, which cover the ear such as shot-blasting helmets or motorcycle helmets, special measurement techniques are required. The methods used are very different from those elsewhere in this appendix where a measurement is made sufficiently far from the head of the exposed person to avoid the disturbed sound field. Measurements very close to the ear are complex and should only be carried out by those with the necessary expertise.

50   There are two techniques for measurements at the ear, a microphone in real ear technique (MIRE) and a manikin technique. MIRE measurements should be performed according to BS EN ISO 11904-1:2002.[22] Measurements using the manikin technique should be performed according to BS EN ISO 11904-2:2004.[23]

MIRE technique

51   This technique is suitable where the sound source itself is not in or near to the ear canal. A miniature or probe microphone is positioned at (or near to) the entrance to the ear canal of the exposed person. During the measurement the sound source should not be displaced from its normal position by the microphone or any measurement accessories. Care also needs to be taken fixing the microphone so it remains in the intended position throughout the measurement and does not come into direct contact with the sound source.

52   BS EN ISO 11904-1:2002 also describes measurements at the eardrum. This guidance does not recommend insertion of a microphone or probe beyond the ear canal entrance.

53   To assess the exposure using criteria applicable to the undisturbed sound field the level measured at the ear is converted to the equivalent free or diffuse field level. This procedure requires analysis of the sound level at the ear into third octave frequency bands; the third octave band levels are then corrected for the frequency response of the microphone and the measurement position. BS EN ISO 11904-1:2002 gives the frequency response of selected measurement positions and also a method for measurement of the microphone and measurement position frequency response. Use of a frequency response from the standard is recommended when the measurement is made at the ear canal entrance.

Manikin technique

54   This technique uses a manikin (sometimes called a head and torso simulator or HATS) fitted with a standard realistic artificial ear. The sound level is measured at the position of the eardrum on the artificial ear. This method allows measurement of sources inserted into the ear canal such as earphones. Care is required with this method to ensure that the source makes contact with the manikin pinna in the same way as on a real ear. Manikin measurements are most suited to laboratory measurements or work activities involving passive listening.

55   The manikin should meet the requirements of clauses 3.4 and 4 in IEC/TR 60959:1990[24] or the equivalent US standard ANSI S3.36:1985.[25] As with the MIRE technique the sound level in the ear is measured in one-third octave bands or narrower bands, and corrected for the manikin frequency response to give the equivalent diffuse or free field level.

56   To obtain the equivalent A-weighted level in a free or diffuse sound field the corrected third octave band levels must be weighted and summed. If unweighted fractional octave band levels have been measured, calculate the A-weighted level in each band by adding the corresponding A-weighting factors. The A-weighted third octave band levels are then summed to give the overall level.

Standards for the performance and testing of noise measurement instruments

Performance

57   The performance of sound level meters, noise dosemeters and sound calibrators is specified by British, European and International standards. Current standards have been produced through the International Electrotechnical Commission and then adopted as European and British Standards. The standards organisations each publish the same standards with the letters BS EN, EN, or IEC prefixing the number according to the publishing organisation. The date following the standard number is when the edition of the standard was adopted by the organisation.

58   New instruments used for noise measurements should meet the current BS EN standards. Instrument standards are subject to review and change, so when you buy new instruments you should ensure they comply with the latest standard. The standards for sound level meters, dosemeters and sound calibrators current at the time of writing are in column 2 of Table 8.

59   Older instruments are not invalidated when standards are superseded. Older instruments that meet the superseded standards in column 3 of Table 8, are also suitable for workplace noise measurements. Instruments meeting a current standard listed in Table 8 may also be used for measurements after the standard is superseded.

60   Sound level meters and sound calibrators are specified to different grades of accuracy. The lower the grading class or type number, the tighter the tolerances placed on the meter's specifications. To maintain the specified tolerances of a sound level meter the sound calibrator class number should be equal to or numerically lower than the class or type number of the meter. All grades of sound level meter are suitable for workplace noise measurements except for Type 3.

Periodic verification

61   The standard procedures for periodic verification of instruments originally manufactured to the instrument standards listed in Table 8 are as follows:

■   **Sound level meters meeting BS EN 61672-1:2003.**[18] At the time of writing this procedure is still in preparation. An interim procedure (Technical Policy Statement TPS 49) has been developed for UKAS (United Kingdom Accreditation Service) accredited testing. See the UKAS website at www.ukas.com.

- **Sound level meters meeting BS EN 60804.**[19] Test to BS 7580:1997[26] parts 1 or 2 as appropriate.

- **Dosemeters.** Procedure included in BS EN 61252:1997.[20]

- **Calibrators.** Verification procedures are included in all versions of BS EN 60942.[21]

62   Using the standard verification procedures, meters and sound calibrators must be tested at least every two years and after any repair likely to affect the performance, to ensure they still meet the standards.

Table 8 Current and superseded standards specifying sound level meters, dosemeters and sound calibrators

Instrument type	Current standards	Superseded standards
Integrating sound level meter	BS EN 61672-1:2003 Also published as IEC 61672-1:2002	BS EN 60804:2001 BS 6698:1986 IEC 804:1985
Dosemeter	BS EN 61252:1997 Also published as IEC 61252:1993 (Previously numbered as IEC 1252:1993 and BS 6402:1994)	
Sound calibrator	BS EN 60942:2003 Also published as IEC 60942:2003	BS EN 60942:1998 IEC 60942:1997

Peak sound pressures

Overview

- What are the typical peak sound pressure levels for common industrial processes?

- How do I select hearing protection for peak sound pressure levels?

- How do I measure peak sound pressure levels?

Risks from high-level peak sound pressures

1   High-level peak sound pressures present a risk to hearing from immediate and permanent hearing loss. The Noise Regulations require employers to take action to reduce the level of exposure if an employee is likely to be exposed to a C-weighted peak sound pressure level (L_{Cpeak}) of 137 dB or above, and place an absolute limit of 140 dB (which can take account of hearing protection).

2   Sources of peak sound pressures may be categorised into the following types:

- Type 1: Low frequency source.

- Type 2: Medium to high frequency source.

- Type 3: High frequency source.

3   Table 9 shows a number of common industrial processes that produce high peak noise levels.

Table 9 Processes producing high peak noise levels

Noise source	Type	Typical C-weighted peak sound pressure level (dB)
Punch press	1	115-140
Jolt squeeze moulding machine	1	120-130
Explosives	1	150-160
Drop hammer	1	130-140
Drop forge	1	130-140
Nail gun/nailer	2	130-140
Hammer (general metalworking)	2	130-150
Proof firing	2	135-140
Rifle fire	2	150-160
Fireworks	2	>140
Pistol	3	140-155
Shotgun	3	150-160

Selecting hearing protection for peak sound pressures

4   When hearing protectors are used in high-level impulsive or impact noise, their attenuation of the peak level can be estimated from the H, M and L values which are supplied with the hearing protectors (see Appendix 3), using a method given in BS EN 458:2004.[16] Note that the methods given in Appendix 3 for estimating attenuation cannot be applied to peak noise.

5   The method uses information on the level of the sound pressure peak together with an assessment of the character of the noise against the three types described above. The information in Table 9 may be used as a guide to peak level and character for the common industrial processes listed. Alternatively, measurements of the C-weighted peak for the source in question may be used. For military impulse sources and industrial impulsive noise sources for which there is no peak level information given in Table 9, specific measurement of the peak level is required.

6   Depending on the character of the noise source, the attenuation provided by a hearing protection device is predicted according to the modified sound attenuation values in Table 10. The effective peak sound pressure level at the ear is estimated by subtracting the modified sound attenuation value from the peak sound pressure level of the impulsive noise source.

Table 10 Sound attenuation values for different impulse or impact noises

Peak noise source type	Modified sound attenuation value (dB)
Type 1	L - 5
Type 2	M - 5
Type 3	H

Example of selection of suitable hearing protection for peak noise

A gamekeeper is exposed to a peak noise of 158 dB from rifle fire. A hearing protection device is required which will reduce the peak level at the ear to at least below the upper exposure action value for peak noise (L_{Cpeak} of 137 dB).

Attenuation required: 21 dB.

Source type: Type 2 Medium to high frequency source.

Decision: Suitable hearing protection will be those devices which have an M-value such that M - 5 is at least 21 dB, ie with a M-value of at least 26 dB.

Other factors influencing selection of hearing protection for peak noise

7   In selecting suitable hearing protection for peak sound pressures, care should be taken that the selected protector is also suitable for the general noise environment, particularly with regard to over-protection. Follow the advice given in Part 5 and Appendix 3.

8   Where the noise exposure is due to high-level discrete impulsive events with quiet periods between, sound restoration earmuffs may be useful. These are designed to provide different attenuation as the sound level changes. Their main purpose is to protect against impulsive or intermittent hazardous noise while allowing communication during quiet periods. These earmuffs reproduce the outside sound under the earmuff cups in quiet conditions. As the sound level increases the gain of the sound-restoration system decreases until the full passive attenuation is achieved.

Measuring peak sound pressures

9   Instruments for measuring high levels of impulsive noise should have an upper limit that extends above 140 dB and an operating range of at least 60 dB to contain the range of levels on a single setting. Meters conforming to BS EN 61672-1:2003[18] Class 2 or better or IEC 60804 Type 1 will be suitable. Meters should be set to measure peak C-weighted levels when measuring peak sound pressures (L_{Cpeak}). Modern meters may have a variety of different time constants, eg Fast (F), Slow (S) and Impulse (I). It is not appropriate to select any of these when measuring peak sound pressures for the purpose of the Noise Regulations. Neither is it appropriate to measure peak pressures using the fast maximum root-mean-square (rms) setting.

10   When measuring peak sound pressures, as with all measurements of noise exposure, the microphone should be positioned close enough to the operator's head to obtain a reliable measure of the noise, but preferably not so close that reflections cause errors. The results are unlikely to be significantly affected by reflected sound if the microphone is kept at least 15 cm away from an operator. The microphone should be placed on the side of the head where the peak levels are higher.

Appendix 3   ## Predicting the attenuation provided by hearing protection

1   Appendix 3 is for anyone who needs to work out what protection will be offered by hearing protectors based on the manufacturer's data. The information here does not apply to the prediction of protection against peak noise, which is covered in Appendix 2 among general information on peak noise.

Hearing protector performance data

2   Suppliers of hearing protection with the CE mark are required to satisfy the relevant part of BS EN 352 which sets out basic safety requirements for hearing protector features such as size, weight and durability for:

- earmuffs (BS EN 352-1:2002);[27]

- earplugs (BS EN 352-2:2002);[28]

- helmet-mounted earmuffs (BS EN 352-3:2002);[29]

- level-dependent earmuffs (BS EN 352-4:2001);[30]

- active-noise-reduction earmuffs (BS EN 352-5:2002);[31]

- earmuffs with electrical audio input (BS EN 352-6:2002);[32]

- level-dependent earplugs (BS EN 352-7:2002).[33]

3   Hearing protection which complies with BS EN 352 must be supplied with performance information derived from a standard test defined in BS EN 13819-2:2002[34] (which in turn draws on a method in BS EN 24869-1:1993).[35] The information required is:

- mean and standard deviation attenuation values at each octave-band centre frequency from 125 Hz to 8 kHz (63 Hz is optional);

- assumed protection values (APV) at each centre frequency (based on mean minus one standard deviation);

- H, M and L values in accordance with BS EN ISO 4869-2:1995;[36]

- SNR value in accordance with BS EN ISO 4869-2:1995.

4   The H, M, L and SNR values are derived from the mean and standard deviation attenuation values.

5   An example of the data supplied by manufacturers is shown in Table 11.

Table 11 Example of the information provided by manufacturers of hearing protection

	Octave band centre frequency (Hz)							
	63	125	250	500	1000	2000	4000	8000
Mean attenuation (dB)	11.5	11.8	11	20.4	22.9	29.8	39.5	39.6
Standard deviation (dB)	4.4	3.2	1.9	3.6	2.2	2.5	2.7	4.9
APV = mean attenuation - std.dev. (dB)	7.1	8.6	9.1	16.8	20.7	27.3	36.8	34.7
Single number values	H	27	M	19	L	13	SNR	22

6   The attenuation data is supplied as a mean and standard deviation to account for the differences in attenuation that people will receive from a particular hearing protector. These differences can occur for a number of reasons, for instance differences in how well a hearing protector fits different people or slight variations in fit each time any one protector is fitted.

7   Non-passive protectors such as level-dependent and active-noise-reduction protectors, and protectors with communication facilities are tested using some additional procedures. BS EN 458:2004[16] includes procedures for the selection of suitable non-passive protectors for a given noise exposure situation. Reference should be made to BS EN 458:2004 when selecting these protectors as the information supplied is not compatible with the procedures given below for predicting the given attenuation.

Methods for predicting the attenuation given by hearing protection

8   The sound pressure levels at the ear when hearing protection is worn may be estimated using a number of different methods. The principal three methods for passive hearing protectors are defined in BS EN 4869-2:1995.

Accounting for 'real world' protection

9   Research has shown that in real use the protection provided can be less than predicted by manufacturer's data. To give a realistic estimate, allowing for the imperfect fitting and condition of hearing protectors in the working environment, it is recommended that a real-world factor of 4 dB is applied. The way this factor is applied is included in the descriptions of the methods that follow.

Table 12 Methods of estimating sound pressure levels using BS EN 4869-2:1995

Method	Description	Data required
Octave-band	Requires detailed data on the frequency content of the noise, and uses information on the attenuation of the protector at specified frequencies.	Octave-band spectrum
HML	Three values H, M and L are used with two simple measurements of the sound pressure level.	A-weighted and C-weighted average sound pressure levels
SNR	The SNR value is used with a single measurement of the sound pressure level.	C-weighted average sound pressure level

10   All methods will give similar predictions of sound levels at the ear for general industrial and occupational noise sources. The HML and SNR methods become less accurate when compared with the octave band method where the noise is dominated by noise at single frequencies, particularly where these are at low frequencies.

11   An electronic spreadsheet to allow the calculations for these methods to be performed is available from the HSE website (www.hse.gov.uk/noise). Paragraphs 12-21 describe the various methods.

Octave-band method

12   The octave-band method is based on an octave-band assessment of the sound pressure level of the noise (this is best done as octave-band values of the unweighted L_{eq}).

13   To calculate the effective A-weighted sound pressure level at the ear (L'_A) using the octave-band method, use the following equation:

$$L'_A = 10 \log \sum_{f = 63(or125)}^{8000} 10^{(L_f + A_f - APV_f)/10}$$

where:

f represents the centre frequency of the octave band in Hz;

L_f is the octave band sound pressure level in octave band f;

A_f is the frequency weighting A for the octave band f; and

APV_f is the assumed protection value of the hearing protector for octave band f.

14   To account for real-world factors, add 4 dB to the calculated L'_A in order to give a more realistic estimate for the protected level at the ear.

HML method

15   The HML method requires measurement of the A-weighted (L_A) and C-weighted (L_C) sound pressure levels. The A-weighted and C-weighted sound pressure levels are used with the three values H (high), M (medium) and L (low) for the protector.

16   The effective A-weighted sound pressure level at the ear, L'_A, is estimated by first calculating the predicted noise level reduction (PNR) afforded by the protector, using one of the two equations below depending on the difference between L_C and L_A for the noise in question.

If $(L_C - L_A) > 2$ dB, then

$$PNR = M - \left[\left(\frac{M - L}{8} \right) (L_C - L_A - 2) \right]$$

otherwise

$$PNR = M - \left[\left(\frac{H - M}{4} \right) (L_C - L_A - 2) \right]$$

17   The PNR should be subtracted from the A-weighted noise level to give the effective A-weighted sound pressure level at the ear (L'_A).

$$L'_A = L_A - PNR.$$

18   To account for real-world factors, add 4 dB to the calculated L'_A in order to give a more realistic estimate for the protected level at the ear.

SNR method

19   The SNR (single number rating) method requires measurement of the C-weighted sound pressure level L_C.

20   The effective A-weighted sound pressure level at the ear, L'_A, is given by subtracting the SNR value for the protector from the C-weighted sound pressure level L_C.

$$L'_A = L_C - SNR.$$

21   To account for real-world factors, add 4 dB to the calculated L'_A to give a realistic estimate for the protected level at the ear.

Appendix 4

Duties of manufacturers and suppliers of machinery

Regulations applying to manufacturers and suppliers of machinery

1   Two main sets of regulations apply to manufacturers and suppliers of tools and machinery where noise is concerned. The Supply of Machinery (Safety) Regulations 1992 as amended[9] (the Supply Regulations) set out the essential health and safety requirements for the design of safe machinery, including detailed requirements for noise. The Noise Emission in the Environment by Equipment for Use Outdoors Regulations 2001[10] set out requirements for declaration of noise levels and, in some cases, achievable levels of noise, for equipment intended for use outdoors.

The Supply of Machinery (Safety) Regulations

2   The Supply Regulations are the United Kingdom's implementation of the European Union's 'Machinery Directive' (Directive 98/37/EC). This Directive exists to support the free movement of goods within the EEA. It establishes minimum health and safety requirements for machinery supplied in the EEA.

Reduction of noise emissions

3   Machinery manufacturers and suppliers are required to design and construct their products so that risks are eliminated, or reduced to a minimum, making it possible for workers to use machinery with the minimum risk to health or safety. It is a particular requirement that risks from noise emissions are reduced to the lowest level taking account of technical progress (Schedule 3, clauses 1.1.2 and 1.5.8).

4   Designers and manufacturers should aim to minimise the noise likely to be generated under all reasonably foreseeable uses of the machine so that people at work are not exposed at levels likely to result in hearing damage. This will involve the application of effective techniques by engineers familiar with noise-control methods.

5   For the risks from noise to be kept to a minimum, noise control needs to be considered at all stages of design and development of the tool or machine.

Provision of information

6   The Supply Regulations require that information on noise emissions must be provided in instructions accompanying machinery, and in technical documents describing the machinery, such as technical sales literature (Schedule 3, clause 1.7.4). The information should:

■   alert users to the noise emission of machines, and help them select suitable machinery and design the work processes for which they will be used; and

■   help users to plan arrangements to protect employees.

7   The information which must be supplied is:

■   a declaration of noise emissions, to include:

❑   the A-weighted sound pressure level at workstations, where this exceeds 70 dB; where this does not exceed 70 dB, this fact must be stated;

❑ the peak C-weighted instantaneous sound pressure value at workstations, where this exceeds 63 Pascals (130 dB);

❑ the A-weighted sound power level emitted by the machinery where the sound pressure level at workstations exceeds 85 dB.

■ any measures needed to keep noise under control when the machine is used (ie instructions for safe use);

■ instructions for assembling and installing the machinery for reducing noise or vibration. For example, a machine may need to be installed on a foundation block, or have anti-vibration mounts fitted, to reduce its noise emission and stop vibration entering the structure of the building, which can cause noise to be radiated from other parts of the building.

8   The supplier must also provide warnings about risks which have not been eliminated and which the user will need to manage, ie 'residual risks' (Schedule 3, clauses 1.1.2 and 1.7.2). This includes, for example, any training requirements for correct use of the machine, instructions for mounting, and any need to restrict the daily duration of use.

Machinery safety standards

9   For many types and classes of machine there are transposed harmonised standards produced by the European Committee for Standardization (CEN) or adopted from the International Organization for Standardization (ISO), setting out safety requirements for the machine in question. Such standards should contain information that allows the designer of the machine to identify potential sources of noise, and to account for these sources to produce a machine that is designed for minimal noise emissions. The transposed harmonised standard should also contain, or provide reference to, a noise test code specifying the operating conditions and measurement methods for measurement of noise emissions.

10   The standard BS EN 1746:1999[37] contains guidance for those drafting machinery safety standards on how to deal with noise. BS EN ISO 12001:1997[38] gives rules for the drafting and presentation of the noise test code, and requires that the operating conditions of the machine specified in the noise test be reproducible and representative of the noisiest operation in typical usage of the machine.

Declared noise emissions

11   In making a declaration of noise emission, it is the supplier's choice whether to follow the methods in the appropriate transposed harmonised standard. If choosing not to follow the standard, or if no standard exists, the supplier must use appropriate measurement methods and operating conditions. In any case, the supplier must indicate the operating conditions and measurement methods used.

12   According to BS EN ISO 4871:1997[39] (the standard which deals with the declaration of noise emission) the declared noise emission should be based on two values:

■ the measured noise emission value (L);

■ the uncertainty of L (K).

13 The supplier should either declare both L and K, or declare a single number with a value equal to $L + K$. In the latter case, it would be good practice for the supplier to declare the K value as well. Incorporating this measure of uncertainty, K, in the declared noise emission allows the supplier to say with some confidence that any tool or machine from the production line of the type to which the declaration refers would give a noise emission value of less than $L + K$ if it was put through the standard test on which the declared values were based.

Residual risks

14 If there is a residual risk, after all practicable means of noise reduction have been incorporated in the design and construction of the machinery, the manufacturer must provide information so that the user can use the machinery safely for all reasonably foreseeable applications. For example, the manufacturers of hand-held grinding machines will need to consider the range of noise levels likely to be generated by their machine when used with various types of abrasive disc and the materials likely to be worked. They can then provide sufficient information to allow their customers to assess and manage the risk when operating the machine.

15 The declared noise emission value will often be enough to alert users to the need to control the noise risk, but where the test code does not produce realistic noise values, additional information is required to allow the equipment to be used safely (eg, by specifying maintenance programmes, operating techniques, training requirements or likely in-use noise levels during the full range of intended uses of the machine).

Training requirements

16 Some tools or machines may require specific training of the operator to ensure that low noise exposures are achieved and sustained. It may also be necessary to train others such as those who will undertake maintenance of machines. Suppliers have a duty to alert users to particular training that is required. For example, this might include:

■ training in new operator skills for tools or machines with noise-reduction features;

■ notification of applications of the tool or machine that produce unusually high noise emissions;

■ information about particular methods of using the tool or machine to be adopted or avoided that greatly affect the emitted noise;

■ training in maintenance requirements to avoid unnecessary exposure.

Presentation of information, labelling and marking

17 All machinery supplied in accordance with the Supply Regulations must be labelled/marked with the following minimum information:

■ the name and address of the manufacturer;

■ CE marking, which includes the year of construction;

■ a designation of series or type;

■ a serial number, if any.

18   The CE mark indicates that the machinery is designed and manufactured to meet all the relevant 'essential health and safety requirements' in Schedule 3 to the Supply Regulations (Annex 1 of the Machinery Directive). These include the duties to minimise risks by design and to provide information on noise emission and the management of residual risks. The CE mark also indicates that the machine complies with any other Directives that apply to the machine. A Declaration of Conformity must accompany the machine; this should state which Directives apply and which standards were followed in its design and manufacture.

19   The information on residual risks should be provided in a way which will be understood easily by the user, for example, using readily understandable pictograms or written information in an appropriate language. Warning information should be included in the instructions accompanying the machine and may also appear in catalogues or in separate data sheets.

Second-hand equipment

20   The Supply Regulations apply to all relevant machinery first supplied or put into service in the EEA from 1993. See paragraphs 260-261 in Part 4 for details of how they apply to second-hand equipment.

Noise Emission in the Environment by Equipment for Use Outdoors Regulations 2001

21   These Regulations relate to equipment that is intended for use in outdoor environments (eg on construction sites or at road works). Types of machinery covered include compressors, pavement breakers, chainsaws and concrete mixers. The requirements under these Regulations are that noise emission (sound power) levels must be marked on all machinery covered by the Regulations (regulation 7(1)). Also, for certain classes of machines covered there are noise limits set in the form of permissible levels. Manufacturers of machines in these classes must ensure that their machinery does not exceed the achievable noise level (regulation 8(1)).

22   Guidance on these Regulations is available from the DTI (www.dti.gov.uk) who administer them.

| Appendix 5 | # Audiometric testing programmes |

1   Appendix 5 advises occupational health professionals and others responsible for audiometric testing on the general approach to carrying out pure tone audiometric testing, methodology, interpreting results and record-keeping.

2   Occupational audiometry is a surveillance technique used to detect early damage to hearing resulting from exposure to noise. Identifying any damage allows appropriate follow-up remedial action in the workplace and any necessary medical referral of the individual. Audiometry is not, in itself, a diagnostic technique, although it can be used to pick up changes in hearing due to many causes, including noise-induced hearing loss (NIHL).

3   Health surveillance is required for all employees regularly exposed above the upper exposure action values and for individuals at greater or additional risk if exposed between the lower and upper exposure action values. Further guidance on to whom the health surveillance requirements apply is in Part 6. This Appendix lays out the recommended procedures for identifying those employees with possible NIHL, to help employers meet their statutory duties. There are other reasons for introducing audiometry, eg considerations of civil liability or fitness for work where good hearing is essential. However, the details of such programmes may differ from those given here.

General approach

4   Before introducing any health surveillance it is important to ensure that employees who will be affected are aware of the implications of the programme. It is helpful to discuss with them and their safety representatives a range of issues including:

- the aims and objectives of the programme;

- the procedures to ensure confidentiality of the results;

- the methods to be followed, including those for medical referral;

- the importance of collating anonymised information for statistical analysis.

Responsibility for the programme

5   There should be a designated person placed in charge of the health surveillance programme. This person should be fully conversant with the technical and ethical aspects of the conduct of occupational audiology, and in particular be responsible for:

- the quality of the service provided;

- ensuring that appropriate standards are maintained during testing;

- record-keeping;

- referring individuals for further advice.

6   A suitable person might be an occupational physician or nurse with specialist training in audiometry, or audiologist. Ultimately, the employer has responsibility for ensuring the proper conduct of health surveillance for noise-exposed employees. It is up to the employer to make clear the responsibilities of the

designated person before the programme is set up. This should include agreement on the detailed arrangements on referral for further advice and the procedures and protocol for feedback of the results to employees, unions and the employer.

7   The person performing the tests may or may not be the same person who is in overall charge of the health surveillance programme. The person actually conducting the tests needs to have, as a minimum, appropriate training so that testing is carried out in a repeatable and accurate manner. A training syllabus for industrial audiometricians has been prepared by the British Society of Audiology (www.thebsa.org.uk) which has approved a number of courses. The basic requirements for any person responsible for conducting the tests are that they should:

- have a good understanding of the aims, objectives and technique of industrial audiometry and how it relates to hearing conservation;

- be competent to ensure an appropriate test environment, to operate and maintain the testing equipment, undertaking basic calibration and to carry out the test procedure;

- understand and comply with the need for confidentiality of personal health information;

- know how to assess and present results according to a defined system and when and how to seek medical assistance.

Pre-test examination

8   It is important that the person being tested has undergone otoscopic examination of the ear immediately before the test to detect any major abnormality or the presence of exudate or wax which might affect the results. The tester should also be familiar with any hearing protection which may be used by workers so they can discuss proper fitting, cleaning and maintenance.

Quality control in audiometry

9   It is important that examinations are made under standardised test conditions with close attention to quality control procedures. Quality control is important to improve the repeatability and reliability of the data produced. Comparisons between test results are an important part of interpretation in an ongoing and effective audiometric programme. All test results therefore need to be comparable by maintaining a standardised method of testing.

10   Careful explanation to the subject of the procedure and familiarisation with the test tones before the test begins are also essential for the collection of reliable data. The criteria used to determine the accuracy with which results are obtained include:

- whether temporary threshold shift is present;

- appropriate and timely equipment calibration; and

- the presence of background noise in the test environment.

Temporary threshold shift (TTS)

11   The best approach to audiometry in relation to the problem of TTS is to seek to eliminate its influence by conducting tests before high exposures to noise occur.

The best method to ensure this is to test individuals before they start work, with advice on reducing noise exposure while travelling to the test. However this will not be practical in most situations. Alternatively it may be useful to advise employees to use additional hearing protection in the period before the test where noise exposure will be present. The aims of this approach are to minimise the influence of TTS and to obtain a measure that is dependent, as far as possible, only on permanent changes to an individual's hearing threshold.

12   Unless there is a prolonged (16 hour or more) period free from high noise levels before testing it is difficult to exclude any contribution from TTS. It is important to ensure that tests are repeated as far as possible in the same conditions from year to year. Be aware that where there are indications of hearing damage needing medical referral, any follow-up will include an audiogram taken free from noise exposure.

Calibration

13   All equipment should be maintained and calibrated according to the recommendations of EN 26189:1991 *Specification for pure tone air conduction threshold audiometry for hearing conservation purposes*[40] and the national standard for audiometers BS EN 60645-1:2001 *Audiometers. Pure-tone audiometers.*[41] In addition to the requirements of this standard it is good practice for the basic calibration to be performed annually. In summary this standard requires that a listening check should be undertaken daily before use and an experienced person with good hearing should listen at each frequency and at three sound intensities to ensure that no extraneous noise is generated by the apparatus. Other checks should be performed weekly and quarterly with a complete overhaul and calibration made annually by a competent laboratory. Many users rely on the manufacturer for this annual check which should incorporate calibration of the earphones with the audiometer. This can be important, as the earphones are often the weakest link in the calibration chain, since they are easily damaged in use.

Test environment

14   EN 26189 gives criteria which should be met in test rooms to prevent test tones being masked by ambient sound levels and to allow measurement of hearing thresholds down to 0 dB. The quietest listening conditions are required at test frequencies of 1 kHz and below. It is usually necessary to use an audiometric soundproof booth to achieve acceptable listening conditions. A small number of people find these claustrophobic and need to be tested outside the booth. Although noise-excluding headsets have been recommended as an alternative method to reducing the effects of ambient noise, variations in fit mean that it is not possible to be certain of the attenuation achieved. This should be considered when comparing results using this strategy. Information should be obtained on the attenuation of the headsets, tested according to BS EN 24869-1:1993 *Acoustics. Hearing protectors. Sound attenuation of hearing protectors. Subjective method of measurement,*[35] which can be used in calculating acceptable background levels.

Methodology

15   Audiometry involves presenting sounds of fixed frequencies and varying intensities to the ear. Testing should follow the methods described in EN 26189 which advises on:

■   the need for otoscopic examination;

■   how to instruct the individual and fit earphones for the test;

- audiometric equipment and its calibration;

- test conditions;

- the detailed conduct of the examination;

- how to determine the hearing threshold level;

- how to construct the audiogram.

Instrumentation

16   There are three main types of audiometer:

- manual;

- self-recording;

- computer-controlled.

17   All have the same function to provide test tones of fixed frequencies and varying intensities at the ears of the person under test. Most people nowadays use self-recording or computer-controlled audiometers. However, manual audiometry may be used where an individual has difficulty co-operating with other techniques. Methods for manual audiometry are included in EN 26189 and the British Society of Audiology has also published advice. There is an increasing reliance on computer-controlled systems, which have the benefit of being able to manipulate results and store and transfer data easily.

Frequency of testing

18   An audiometric programme should consist of a baseline audiogram conducted before employment where noise is a hazard, followed by a schedule of audiometric testing to monitor hearing threshold levels following exposure to noise at work. For quality control purposes it is particularly important to obtain a baseline that, as far as possible, is not contaminated by TTS. This reflects the importance of this initial test as a reference point for all future comparisons.

19   The schedule of audiometric testing should include annual tests for the first two years of employment and at three yearly intervals thereafter. More frequent testing may be required if significant changes in hearing level are detected or exposure conditions change, increasing the risk of hearing damage. As a quality control measure, it would be prudent to repeat any audiogram which showed a difference from the previous result of more than 10 dB at any frequency.

20   At the baseline examination it is important to obtain information about the individual's job, previous noise exposures and medical history (see example questionnaire in Appendix 6). At all subsequent tests the individual should be asked about any changes in personal circumstances, work patterns and noise exposure, and any complaints relating to the ears or hearing. If changes are indicated, previous records should be revisited and amended as necessary.

21   Where a workforce is already exposed to noise before the audiometric programme begins, the baseline audiogram will simply be the first test to be made. If there is no evidence of hearing loss, subsequent testing can follow the suggested schedule. Where damage is detected at the baseline, actions taken should follow the advice given in paragraphs 31-37 in this appendix.

Interpretation of results

22   The initial assessment of an audiogram will normally be made by the person conducting the test, or by a medical practitioner. The results of previous audiograms should be available for comparison. The tester should then consider:

■   whether any immediate action is required;

■   what information the audiogram gives about the change in hearing level and its rate of progression.

23   To help with the initial assessment and interpretation of the audiogram and to guide the tester as to the appropriate action to take, a categorisation scheme has been developed by HSE (see Table 13). This scheme replaces that previously endorsed in HSE guidance note MS26. It is recommended that this or a similar scheme should be used for the initial assessment of audiograms. The scheme provides a guide for action which can be adapted in light of local experience.

24   In this scheme, the criteria for audiometric classification are based on a summation of the hearing levels obtained at 1, 2, 3, 4 and 6 kHz. This calculation should be done for each ear separately. This sum of frequencies has been chosen as being representative of the effects of NIHL. Although this scheme recommends a sum of hearing levels at specific frequencies, it is important that audiometry is still conducted at 0.5 and 8 kHz.

Method for evaluating audiograms

25   Once the test has been completed, the relevant quality control issues have been taken into consideration and a noise and health questionnaire completed, the following steps should be carried out to categorise the audiogram. Each category has a descriptor relating to the condition of an individual's hearing and advises what steps should be taken next. Table 13 provides details of the four categories.

26   Firstly you should add up the hearing levels obtained at the 1, 2, 3, 4 and 6 kHz frequencies so that a single value is obtained for each ear. Table 14 provides the relevant warning and referral thresholds for these sums taking into account the age and gender of the individual.

■   **If the sum for both ears is below the warning level** then that individual will fall within **category 1 – acceptable hearing ability.**

■   **If the sum for either ear exceeds or is equal to the warning threshold level** for their respective age and gender then the individual will fall into **category 2 – mild hearing impairment.**

■   **If the sum exceeds or is equal to the referral level** for either ear then the individual would fall into **category 3 – poor hearing** and would require referral for further medical advice.

27   To determine whether there has been a rapid loss in hearing since the last examination a sum of the hearing thresholds obtained at 3, 4 and 6 kHz should be made. If the previous test was performed within the last three years and an **increase in hearing threshold level of 30 dB or more (as a sum of 3, 4, and 6 kHz)** is found then this individual would fall into **category 4 – rapid hearing loss** and require referral on for further medical advice.

28   Following this logical method should result in each individual being placed into one of the four categories.

29   A further sum should be undertaken to determine whether the individual has any unilateral hearing loss suggesting a problem due to disease or infection. Sum the hearing levels at 1, 2, 3 and 4 kHz for both ears. If the **difference between the ears is greater than 40 dB** the individual should be advised of the findings and referred on for further medical advice.

30   Interpretation of an audiogram may highlight effects other than NIHL. Further tests will be required to ascertain the causes of any abnormal audiogram. These will be conducted by trained medical professionals following referral.

Table 13 The HSE categorisation scheme

Category	Calculation	Action
1 ACCEPTABLE HEARING ABILITY Hearing within normal limits.	Sum of hearing levels at 1, 2, 3, 4 and 6 kHz.	None
2 MILD HEARING IMPAIRMENT Hearing within 20th percentile, ie hearing level normally experienced by 1 person in 5. May indicate developing NIHL.	Sum of hearing levels at 1, 2, 3, 4 and 6 kHz. Compare value with figure given for appropriate age band and gender in Table 14.	Warning
3 POOR HEARING Hearing within 5th percentile, ie hearing level normally experienced by 1 person in 20. Suggests significant NIHL.	Sum of hearing levels at 1, 2, 3, 4 and 6 kHz. Compare value with figure given for appropriate age band and gender in Table 14.	Referral
4 RAPID HEARING LOSS Reduction in hearing level of 30 dB or more, within 3 years or less. Such a change could be caused by noise exposure or disease.	Sum of hearing levels at 3, 4 and 6 kHz.	Referral

Actions/Advice

31   All individuals should be given advice regarding the effect of noise on hearing and the correct use of hearing protectors as part of the health surveillance programme (see Appendix 6 for examples).

32   Where the individual falls **within category 2 a formal notification** should be given to that employee regarding the presence of hearing damage. This should include reference to the extent and implication of the damage and ways in which to minimise or prevent any further damage or loss. Retraining and reinforcement of the correct use of hearing protection and the importance of complying with other hearing conservation methods provided by the employer are the main points to stress. It is recommended that this information be given verbally, while being supported by written documentation for future reference. An example of a formal warning to those categorised as having mild hearing damage is in Appendix 6. It is also good practice to provide the employee with a copy of their audiogram following each test.

33   Arrangements and procedures should be put in place for **medical referral** of those individuals falling into **categories 3 and 4** and where unilateral hearing loss is identified. Where referral is indicated, audiograms should first be brought to the attention of a medical practitioner, which could be the occupational physician involved with the health surveillance programme or otherwise the employee's general practitioner or to an audiologist where available. In some cases the further advice of a consultant ear, nose and throat surgeon will be required. An example letter of referral is in Appendix 6.

Table 14 Classification of audiograms into warning and referral levels

| Age | Sum of hearing levels 1, 2, 3, 4 and 6 kHz | | | |
| | Males | | Females | |
	Warning level	Referral level	Warning level	Referral level
18-24	51	95	46	78
25-29	67	113	55	91
30-34	82	132	63	105
35-39	100	154	71	119
40-44	121	183	80	134
45-49	142	211	93	153
50-54	165	240	111	176
55-59	190	269	131	204
60-64	217	296	157	235
65	235	311	175	255

34   If individuals fall into category 4, the frequency of testing should be reconsidered and will need to be more frequent than at three-yearly intervals.

Further assessment of audiograms

35   For those audiograms in categories 1 and 2, it is desirable to make an assessment of the rate of progression of any changes due to noise exposure. This is to provide an early warning of damage in cases where hearing loss develops at a rate greater than might be expected due to age and gender, but where the referral criteria have not been reached. The tester will need to have the knowledge and experience necessary to detect the 'normal' changes to be expected between tests.

Other actions

36   Where audiometric results have not triggered a referral, but it is clear that hearing loss has become a handicap to the individual it may be appropriate to consider referral. This hearing loss may not indicate anything other than normal ageing. Referral to a medical practitioner will enable a full examination to determine whether provision of a hearing aid may be of benefit. Medical referral is also appropriate where an individual reports symptoms such as ear pain, discharge, dizziness, severe or persistent tinnitus, fluctuating hearing impairment or a feeling of fullness or discomfort in one or both ears. These problems may be determined through a noise and health questionnaire or other form of interview at the test.

37   Where there is concern about changes in hearing thresholds, or where exposure conditions have altered, a repeat audiogram may be requested before the next scheduled routine test.

Record-keeping

38   The person in charge of the audiometric programme should maintain records of the programme, including:

■   any questionnaires completed;

■   the audiograms themselves;

■   any assessments made of the results.

39   An employer's occupational health department will usually store the records. Otherwise the operator who conducted the tests can be responsible for record-keeping on behalf of the employer. Where a visiting service conducts the testing, the results should be maintained by the employer, separate from personnel records. A person should be nominated to keep the records and to oversee access to them. Original records should be kept up to date as long as the employee remains under health surveillance. Employers should keep the health record as long as an individual remains in their employment, and may wish to retain it for longer as enquiries regarding the state of an individual's hearing may arise many years after exposure to noise has ceased.

40   The degree of confidentiality of the audiometric results should be agreed between the employer and employees and their representatives before testing begins. This will determine what information about an individual may be released to a third party without their written consent. Audiometric results and noise and health questionnaires would be considered medical in confidence. Normal professional ethics require consent to be obtained before passing results of testing to the employee's general practitioner or anyone else.

41   Employers will need to see anonymised grouped data on the hearing of the workforce to advise them of the effectiveness of their noise-control measures. This can be done in a way that does not reveal details of any particular individual's hearing threshold and does not compromise the issue of confidentiality. Consent will not be required for this type of information to be provided to the employer.

42   Figure 43 provides an outline of the procedures to follow in an audiometric testing programme.

Future work in noisy environments

43   Following referral, if noise-induced hearing loss is deemed to be stable, continuing exposure to noise will usually be acceptable where adequate hearing protection is used and where residual hearing ability is not so poor as to make the risk of further hearing loss unacceptable. In exceptional circumstances a medical professional may indicate to the employer that an individual is no longer fit for their current employment.

44   There will be a few employees who have responsibility for the safety of others and who need to communicate easily and to hear auditory warning signals. Here, severe hearing loss will cause difficulties and audiometric testing may be used as an assessment for fitness for work. The requirement to maintain good hearing should be made clear at the time of recruitment. Audiometric assessment based on the average hearing threshold over the speech frequencies or those of particular warning sounds may be helpful, but are not detailed here.

Figure 43 Flow diagram for audiometric testing and categorisation

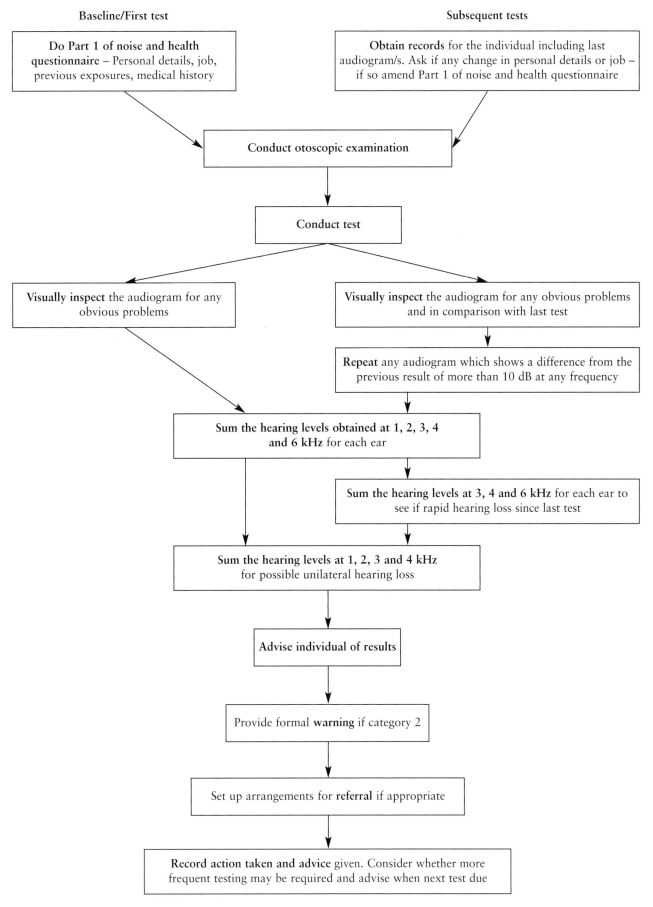

Baseline/First test

Subsequent tests

Do Part 1 of noise and health questionnaire – Personal details, job, previous exposures, medical history

Obtain records for the individual including last audiogram/s. Ask if any change in personal details or job – if so amend Part 1 of noise and health questionnaire

Conduct otoscopic examination

Conduct test

Visually inspect the audiogram for any obvious problems

Visually inspect the audiogram for any obvious problems and in comparison with last test

Repeat any audiogram which shows a difference from the previous result of more than 10 dB at any frequency

Sum the hearing levels obtained at 1, 2, 3, 4 and 6 kHz for each ear

Sum the hearing levels at 3, 4 and 6 kHz for each ear to see if rapid hearing loss since last test

Sum the hearing levels at 1, 2, 3 and 4 kHz for possible unilateral hearing loss

Advise individual of results

Provide formal **warning** if category 2

Set up arrangements for **referral** if appropriate

Record action taken and advice given. Consider whether more frequent testing may be required and advise when next test due

Effectiveness of the hearing conservation programme – information to be passed to the employer

45   Grouped, anonymised analysis of audiometric results can provide useful information to the employer about the overall effectiveness of a hearing conservation programme. This is essential as it ensures the results of the testing are acted upon when required. The analysis can be a simple tabular presentation of the percentage of workers falling into each category compared with previous test results (as long as there has not been a significant change in the work population). This may be broken down for different groups of workers or different areas of the workplace. The use of computerised audiometer systems often facilitates these type of statistical assessments, as they can be pre-programmed to provide such information. The form of assessment which is appropriate often depends on the number of workers exposed to noise. This type of anonymised interpretation of the results does not compromise confidentiality.

46   Where analysis of the audiometric results demonstrates that there has been a deterioration in hearing, perhaps in particular groups of workers, a reassessment of the exposure factors will be required. The results may reflect a change in exposure conditions, for example as a result of relocation of machinery, changes to working patterns, inadequate maintenance of machinery or a failure of hearing conservation methods, in particular failure or inefficient use of hearing protection.

Appendix 6

Sample documents

Noise and health questionnaire

PART 1 to be completed at baseline test and amended as necessary

Personal details

Surname:

Forename:

Sex: MALE/FEMALE

Date of birth: /   /

Age:

Home address:

Employment

Present employer:

Location/department:

Job title/occupation:

Clock/staff no:

Time in this post:

Medical history

Do you consider your hearing to be:	Left ear	Good/Fair/Poor
	Right ear	Good/Fair/Poor
Do you wear a hearing aid?		YES/NO
Have you suffered any injury/trauma to your ears? If so describe:		YES/NO
Earache, discharging ears or other ear disease as child or adult? If so describe:		YES/NO
Any ear disease or deafness in the family? If so describe:		YES/NO

Ever suffered head injury/concussion/unconsciousness?   YES/NO
If so describe:

Do you suffer ringing in the ear/head?   YES/NO

Do you suffer from dizziness/giddiness?   YES/NO

Exposure to ototoxic drugs or solvents?   YES/NO
eg streptomycin, otosporin, quinine, toluene

Exposure to gunfire/blasts/explosions?   YES/NO
If so describe:

Do you have any noisy hobbies?   YES/NO
Tick all that apply: Motor sports
                       Ride a motorcycle
                       DIY
                       Discos/loud music
                       Shooting
                       Other

Do you hear better or worse in noise?   BETTER/WORSE

Have you had wax removed from your ears?   YES/NO
If YES, when?

Previous noise exposure   Include past noisy jobs where you have had to shout to be heard

Previous job: For how long? Yrs mths

Ear protection provided?   YES/NO

Type: Glass wool/earplugs or inserts/earmuffs

Worn?   YES/NO

Previous job: For how long? Yrs mths

Ear protection provided?   YES/NO

Type: Glass wool/earplugs or inserts/earmuffs

Worn?   YES/NO

Previous job: For how long? Yrs mths

Ear protection provided?   YES/NO

Type: Glass wool/earplugs or inserts/earmuffs

Worn?   YES/NO

Previous job: For how long? Yrs mths

Ear protection provided?   YES/NO

Type: Glass wool/earplugs or inserts/earmuffs

Worn?   YES/NO

Previous job: For how long? Yrs mths

Ear protection provided?   YES/NO

Type: Glass wool/earplugs or inserts/earmuffs

Worn?   YES/NO

PART 2 to be conducted at each test

Date of test: /   /

Otoscopic examination

Wax in auditory canal?   LEFT RIGHT
(ie <50% of tympanic
membrane visible)

Exudate in auditory canal?   LEFT RIGHT

Tympanic membrane   normal/scarred/perforated   LEFT

                                  normal/scarred/perforated   RIGHT

Other comments:

TTS effects

Exposure to noise in past 48 hrs?   NO   YES   Specify:

Hearing protection worn before test?   NO   YES   Specify:

Results

Category   1 2 3 4   (circle as appropriate)

                Unilateral hearing loss

Any other problem identified which requires referral?

Actions

Advice given:

Official warning given? (Category 2)   YES/NO

Referral for further medical advice set up? (Categories 3 and 4)   YES/NO

Date of next test   /   /

Signed by:
(person conducting test)

Signed by employee:

Example 1

Example of general advice for employees undergoing health surveillance for NIHL

Note: Supplementing this with a demonstration of how to fit the relevant hearing protection correctly would be useful.

Effects of noise on hearing

Hearing ability deteriorates with age. However, exposure to high levels of noise at work or through hobbies and leisure activities over time will cause irreparable damage to hearing. Therefore high noise exposures are likely to cause deafness at an earlier age than would be expected naturally.

You may only realise the extent of your hearing loss when it has become so bad that your family complain that you have the television too loud, or you realise you cannot keep up with conversations. This permanent hearing loss is incurable and young people can be damaged as easily as the old.

So what can you do about it?

Hearing loss is permanent and irreversible. However noise-induced hearing loss is completely preventable. Your employer has put in place various systems to reduce the amount of harmful noise in your workplace. You should be aware of these systems and comply with them at all times. This may mean not entering a room or work area, keeping doors/shields/guards in place or in some cases, where damaging noise cannot be reduced, you will be required to wear hearing protection such as earplugs or earmuffs. Areas where hearing protection is required will be clearly marked with signs.

Your employer has provided training on how to wear your hearing protection correctly. It will only work if used properly. It is your duty, and within your own interests to protect your hearing to wear the protection correctly at all times.

Please also be aware that any hobbies you have or leisure activities which involve noise to a level that you find yourself having to shout above are likely to be harmful. Some examples are riding motorbikes, shooting or listening to loud music (concerts/pubs/clubs). These types of noise are just as harmful as those at work and will affect your hearing in the same way. You can therefore protect your own hearing by reducing your exposure to such harmful levels of noise outside work.

Example 2

Example employee 'mild hearing impairment' notification

The results of your audiometric test today have indicated that you have a 'mild hearing impairment' compared to other men/women of your age group. This could be due to noise exposure at work or outside work due to your hobbies, involvement in noisy activities or other lifestyle habits.

This type of hearing damage is irreversible, so it won't get better and what you lose you don't get back. Hearing deteriorates in everybody with age so older people have less hearing ability than the young. However, this type of damage is in addition to the general hearing loss over time and so you may become more deaf earlier than other people your age.

We have identified this small amount of damage early due to the ongoing testing of your hearing. This is a warning to let you know that if you continue to be exposed to high noise levels more irreversible damage is likely to occur.

As we have identified this damage early, before the need to refer you for medical help, you can help prevent any further deterioration by:

- being more vigilant in wearing your hearing protection and wearing it properly;

- ensuring you comply with any other hearing conservation procedures in your workplace;

- reducing your exposure to excessive noise in your hobbies and leisure activities.

Example 3

Example employee referral notifications

Poor hearing

The results of your audiometric test today have indicated that you may have 'poor hearing' which may be due to excessive exposure to noise either at work or in your hobbies or leisure activities. This means your hearing is worse than would be expected normally for your age. I am referring you to your GP for further investigation of the extent and possible causes of this damage. Please take note of the advice given to you today, as it is now more important that you conserve what hearing ability you still have. We will continue to monitor your hearing through this programme.

Rapid hearing loss

The results of your audiometric test today have indicated that you have demonstrated a 'rapid hearing loss' since your last test. This rate of loss is significant and therefore requires that you be referred to your GP for further investigation and clarification of the extent of hearing damage and possible cause of this. Your GP may also be able to provide you with a hearing aid if this is appropriate. Please take note of the advice given to you today, as it is now more important that you conserve what hearing ability you still have. We will continue to monitor your hearing through this programme.

Unilateral hearing loss

The results of your audiometric test today have indicated that you have a 'unilateral hearing loss', which means that your hearing in one ear is much worse than the other. This is not usually due to noise but your GP will be able to investigate in more detail. I am therefore referring you to your GP for further advice and treatment as appropriate. We will continue to monitor your hearing through this programme.

Example 4

Example referral letter to a General Practitioner

Date

Dear [insert GP's name]

RE: (Employee's name, date of birth, address)

I understand that [employee's name] is a patient of yours. I am referring him/her to you following a routine audiometric test as part of his employment at [company].

This person is exposed to potentially harmful noise at work. His/her most recent audiogram, performed on [date], has demonstrated that he/she has poor hearing/rapid hearing loss since his or her last test/unilateral hearing loss/[other problems with the ear or hearing which should be detailed]. The relevant audiograms are enclosed for your information.

I would be grateful if you could examine/arrange for [employee's name] to be examined to ascertain the cause of the hearing loss and provide the necessary treatment.

An opinion is also sought as to whether his/her hearing condition is consistent with noise-induced hearing loss, stable, and likely to be aggravated by further exposure to noise. If you require any further information please contact me on/at [contact details] quoting [reference details].

Yours sincerely

[Name, Title]

References

1   *Noise at work: Guidance for employers on the Control of Noise at Work Regulations 2005* Leaflet INDG362(rev1) HSE Books 2005 (single copy free or priced packs of 10 ISBN 0 7176 6165 2)

2   *Protect your hearing or lose it!* Pocket card INDG363(rev1) HSE Books 2005 (single copy free or priced packs of 25 ISBN 0 7176 6166 0)

3   *Managing health and safety in construction: Construction (Design and Management) Regulations 1994. Approved Code of Practice and guidance* HSG224 HSE Books 2001 ISBN 0 7176 2139 1

4   *The Health and Safety (Training for Employment) Regulations 1990* SI 1990/1380 The Stationery Office 1990 ISBN 0 11 004380 4

5   *Management of health and safety at work. Management of Health and Safety at Work Regulations 1999. Approved Code of Practice and guidance* L21 (Second edition) HSE Books 2000 ISBN 0 7176 2488 9

6   *Safety representatives and safety committees* L87 (Third edition) HSE Books 1996 ISBN 0 7176 1220 1

7   *A guide to the Offshore Installations (Safety Representatives and Safety Committees) Regulations 1989. Guidance on Regulations* L110 (Second edition) HSE Books 1998 ISBN 0 7176 1549 9

8   *A guide to the Health and Safety (Consultation with Employees) Regulations 1996. Guidance on Regulations* L95 HSE Books 1996 ISBN 0 7176 1234 1

9   *Machinery. Guidance notes on UK Regulations. Guidance on the Supply of Machinery (Safety) Regulations 1992 as amended by the Supply of Machinery (Safety) (Amendment) Regulations 1994* URN 95/650 Department of Trade and Industry 1995 (Available from the DTI Publications Orderline: 0845 015 0010)

10   *The Noise Emission in the Environment by Equipment for Use Outdoors Regulations 2001* The Stationery Office 2001 SI 2001/1701 ISBN 0 11 029485 8 amended by *The Noise Emission in the Environment by Equipment for Use Outdoors (Amendment) Regulations 2001* SI 2001/3958 ISBN 0 11 039010 5

11   *Safe use of work equipment. Provision and Use of Work Equipment Regulations 1998. Approved Code of Practice and guidance* L22 (Second edition) HSE Books 1998 ISBN 0 7176 1626 6

12   BS ISO 230-5:2000 *Test code for machine tools. Determination of the noise emission*

13   ISO 7960:1995 *Airborne noise emitted by machine tools. Operating conditions for woodworking machines*

14   BS EN ISO 9902-1:2001 *Textile machinery. Noise test code. Common requirements*

15   BS EN 1265:1999 *Noise test code for foundry machines and equipment*

16   BS EN 458:2004 *Hearing protectors. Recommendations for selection, use, care and maintenance. Guidance document*

17   *The Personal Protective Equipment Regulations 2002* SI 2002/1144 The Stationery Office 2002 ISBN 0 11 039830 0

18   BS EN 61672-1:2003 *Electroacoustics. Sound level meters. Specifications*

19   BS EN 60804:2001 *Integrating-averaging sound level meters (superseded, withdrawn)*

20   BS EN 61252:1997 *Electroacoustics. Specifications for personal sound exposure meters*

21   BS EN 60942:2003 *Electroacoustics. Sound calibrators*

22   BS EN ISO 11904-1:2002 *Acoustics. Determination of sound immission from sound sources placed close to the ear. Technique using a microphone in a real ear (MIRE technique)*

23   BS EN ISO 11904-2:2004 *Acoustics. Determination of sound immission from sound sources placed close to the ear. Technique using a manikin*

24   IEC/TR 60959:1990 *Provisional head and torso simulator for acoustic measurements on air conduction hearing aids*

25   US standard ANSI S3.36:1985 *Specification for manikin for simulated in situ airborne acoustic measurements*

26   BS 7580:1997 *Specification for the verification of sound level meters*

27   BS EN 352-1:2002 *Hearing protectors. Safety requirements and testing. Ear-muffs*

28   BS EN 352-2:2002 *Hearing protectors. Safety requirements and testing. Ear-plugs*

29   BS EN 352-3:2002 *Hearing protectors. Safety requirements and testing. Ear-muffs attached to an industrial safety helmet*

30   BS EN 352-4:2001 *Hearing protectors. Safety requirements and testing. Level-dependent ear-muffs*

31   BS EN 352-5:2002 *Hearing protectors. Safety requirements and testing. Active noise reduction ear-muffs*

32   BS EN 352-6:2002 *Hearing protectors. Safety requirements and testing. Ear-muffs with electrical audio input*

33   BS EN 352-7:2002 *Hearing protectors. Safety requirements and testing. Level-dependent ear-plugs*

34   BS EN 13819-2:2002 *Hearing protectors. Testing. Acoustic test methods*

35   BS EN 24869-1:1993 *Acoustics. Hearing protectors. Sound attenuation of hearing protectors. Subjective method of measurement*

36   BS EN ISO 4869-2:1995 *Acoustics. Hearing protectors. Estimation of effective A-weighted sound pressure levels when hearing protectors are worn*

37   BS EN 1746:1999 *Safety of machinery. Guidance for the drafting of the noise clauses of safety standards*

38   BS EN ISO 12001:1997 *Acoustics. Noise emitted by machinery and equipment. Rules for the drafting and presentation of a noise test code*

39   BS EN ISO 4871:1997 *Acoustics. Declaration and verification of noise emission values of machinery and equipment*

40   EN 26189:1991 *Specification for pure tone air conduction threshold audiometry for hearing conservation purposes*

41   BS EN 60645-1:2001 *Audiometers. Pure-tone audiometers*

Further information

HSE priced and free publications are available by mail order from
HSE Books, PO Box 1999, Sudbury, Suffolk CO10 2WA Tel: 01787 881165
Fax: 01787 313995 Website: www.hsebooks.co.uk (HSE priced publications are
also available from bookshops and free leaflets can be downloaded from HSE's
website: www.hse.gov.uk.)

British Standards are available from BSI Customer Services,
389 Chiswick High Road, London W4 4AL Tel: 020 8996 9001
Fax: 020 8996 7001 e-mail: cservices@bsi-global.com
Website: www.bsi-global.com

The Stationery Office publications are available from The Stationery Office,
PO Box 29, Norwich NR3 1GN Tel: 0870 600 5522 Fax: 0870 600 5533
e-mail: customer.services@tso.co.uk Website: www.tso.co.uk (They are also
available from bookshops.)

For information about health and safety ring HSE's Infoline
Tel: 0845 345 0055 Fax: 0845 408 9566 Textphone: 0845 408 9577
e-mail: hse.infoline@natbrit.com or write to HSE Information Services,
Caerphilly Business Park, Caerphilly CF83 3GG.

Printed and published by the Health and Safety Executive   C100   10/05